THE MG LOG

THE  LOG

Edited by
PETER HAINING

Editorial Consultant
JEAN COOK

SOUVENIR PRESS

First published 1993 by Souvenir Press Ltd,
43 Great Russell Street, London WC1B 3PA
and simultaneously in Canada

ISBN 0 285 63144 6

Photoset by Rowland Phototypesetting Ltd,
Bury St Edmunds, Suffolk

Printed in Great Britain by
Mackays of Chatham PLC, Chatham, Kent

Contents

Acknowledgements

I should like to acknowledge my gratitude to the following people who have provided text and illustrations for this book: Bob Pattrick, Enid Wilkinson, Jean Cook, Cyril Mellor, Barry Foster, Anna Guerrier, David Saunders, Geoff Shirt, Mike Allison, Phil Jennings, Don Hayter, Richard Monk, Ron Cover and Richard Morris. My thanks, too, to HarperCollins for permission to quote from Sir Michael Edwardes' book *Back from the Brink* (1983) and Random Century Publishing Group for the extract from *Where The Rivers Run* by John Wain (1988); also to Greg Allport and Rover Cars for permission to reproduce the MG logo. My special thanks also go to Jean Cook, without whom this book would not have been possible, Jean's husband, Dennis Cook, and my wife Philippa and daughter Gemma, both of whom have also fallen under the spell of 'MG Magic'.

Peter Haining
April 1993

Facing page: The archetypal MG, a 1949 TC, 'the Sports Car America Loved First', at speed. (The RoadRunner Collection)

Preface

An advertisement which appeared in newspapers, magazines and on roadside hoardings not so many years ago declared: 'At no extra cost, the MGB gives you the sky, the sun and the wind'. The poster actually first appeared in 1963 to promote what was to become the most successful of all MG models—but could just as easily have referred to the very earliest sports cars which Cecil Kimber, the founder of the MG empire, began to produce at Morris Garages in Oxford in 1923. What Kimber did was to revolutionise the motor industry by introducing for the first time a two-seater 'sporting' car which contrasted vividly with the staid and sober four-seater family saloons of the period.

Kimber was a man with a vision. He sensed that beneath the British reserve

lay a yearning for motoring that was inexpensive and *fun*. When a series of distinctive little cars with their red octagon on the radiator began to appear on the roads of England in the late Twenties, a legend was born—a legend that has now spread to the far corners of the Earth. Wherever, that is, people enjoy driving cars of character, made for individuals.

The legend of the MG is one that hardly anyone who has driven a car can be unaware of: a nostalgia for a unique marque among so much motoring conformity. But it is not just a story about a car—because it takes people to make cars, and those who have made and driven the various models of the MG have left the marks of their personalities upon each one. This book is about those people and the cars they have made so special.

It is a book that has grown out of many happy hours spent in conversation with the one person closest to the heart of the legend, Jean Cook, Cecil Kimber's daughter, whose judgement of her father as a man and a mechanical genius is both intimate and objective. Though she does not claim to be an expert on the technical details of the MG, there is little she does not know about the people who have been associated with the car . . . both yesterday and today. Thanks to her, a group of the best known MG enthusiasts have been brought together in the pages of this book to give their impressions of 70 years of Octagon history. It is a story that offers some surprises, too, such as refuting the legend that the MG known as 'Old Number One' was actually Kimber's first car, as well as explaining exactly what the initials MG stand for . . .

Like many of the contributors, Jean has dipped into her family archives and photograph albums to provide a wealth of special pictures, many of which have not appeared previously in book form, and which will doubtless excite the specialist as much as the layman. These illustrations, plus the stories about the men—and women—who were a part of motoring history, provide what I believe to be a unique insight into the legend of the MG— the sports car *everyone* loves.

The great ambassador of the MG world, Jean Cook, Cecil Kimber's daughter. (Peter Jenniches)

Octagonal Fever

Peter Haining

Most families have a story to tell of 'Octagonal Fever'. Stories of owning a particular MG model, experiences of driving one, or simply just wishing they had one—because there has never been a sports car quite like this particular marque, with its distinctive appearance and hint of the open road and motoring as it really should be. I am no exception, which is one of the reasons I have assembled this celebration of 70 years of a unique motor car.

My story may be unexceptional, but I believe it *is* typical of a great many others. It also illustrates why the MG lives on in the memory when far more exciting, even dramatic or dangerous, experiences in other cars have long been forgotten. The year was 1954, the month August, and I was a year into my teens when my uncle, Bob Pattrick, then serving in the RAF, called at our home in Loughton, Essex. He pulled up outside in this stylish and throaty little sports car which I now know as an MG J1 Midget, which had been first registered in 1933 and which he had bought in May 1954. With its top down and a battery of three huge headlights around the distinctive radiator grille, FS 5525 was the stuff of which young boys' dreams are made.

Bob was stationed at RAF Uxbridge and was going home on leave to Scarborough in Yorkshire. He was breaking his drive in Loughton to see his older sister—my mother—and catch up on any family news to relate to their mother, my grandmother. I have no idea what they talked about as they sipped tea in the garden on that summer day, because all the time I was sitting in a kind of trance behind the wheel of the Midget, imagining driving furiously around Silverstone Race Track.

When my uncle came out and asked if I would like to drive with him up to Yorkshire to spend some of my school holidays with my grandmother, I'm not sure I heard him the first time. When my father, standing beside him, repeated the offer and the truth began to sink in that I could spend the next few hours motoring in this amazing machine, I couldn't agree fast enough.

9

The J1 in which the author had his fateful London to Yorkshire run.

A case was quickly packed for me, and within the hour we were on the A11, bound for the north.

With the rumble of the engine just ahead of our feet, and the wind whistling around our ears, it was not easy to make conversation. But I was still lost in my dreams as we motored past lorries, coaches and sedate family saloons. Even when the journey took a turn for the worse, nothing could disturb my youthful equanimity—and, indeed, I subsequently had to be reminded of the problems we encountered.

The first was rain, which broke over us as we were driving through Lincoln-shire. Bob had to pull over to the side of the road and put the hood up. I'm sure I would have been happy to sit in the rain, but he valued the car and did not want the leather upholstery spoiled in a downpour. As we motored on, I suddenly became aware that my legs, stretched out in front of me, were getting wet.

Bob, of course, knew what was happening, and although I made no protest, he explained to me. 'In these "J" types your legs are only separated from the road by the sheet metal of the bonnet,' he said. 'I'm afraid it's not exactly water-tight. The water is coming up from the puddles on the road. That's one of the problems with an MG—wet legs are always a hazard.'

My uncle says now that I was not at all amused about getting wet—but I think he exaggerates. What boy doesn't mind a little discomfort as the cost of a genuine adventure?

Further on, as we neared the Yorkshire border, a burst of blue smoke suddenly entered the cockpit. I wondered what on earth had happened. In fact, Bob went on driving as if nothing were wrong, and we reached Scarborough an hour or so later without mishap. It was to be some months before he explained what had occurred.

'When I saw the smoke I realised that this was the first sign of serious piston trouble,' he said. 'In fact, the car was sound enough to make the return journey to Uxbridge, where I spent a couple of weeks sorting out the trouble.'

My uncle, as you may have gathered, was—and is—a mechanically minded man. But the one driving force he couldn't master in furthering his association with that MG was . . . love. The year after our drive, he met his future wife, Sheila, who he says was no more keen on getting her legs wet than I was (!), and he decided to sell the J1 to an RAF colleague in favour of a sports saloon. It was to be quite a few years—not to mention the raising of two sons—before Bob recommenced his love affair with the marque by buying a rubber-bumpered Midget, RFW 159W, in 1980.

How Bob got his admiration for the MG also curiously parallels my own. It started in the 1930s when another of his sisters, my Aunt Enid, was being courted by an MG enthusiast named Philip White, who was the proud possessor of what he told Bob was a Midget J4. Young Bob, who was as smitten with that car—UG 4844—as I had been with his J1, had no reason at the time to doubt the statement. It was not until some years later—and after Philip White had been tragically killed in a German air raid on Brighton in 1943—that he discovered that the J4 was actually a J2. A photograph of the car which my aunt has kept (and which is reproduced here) helped him to solve the mystery, as he explains:

Philip White's J2 which he mistakenly believed to be a J4.

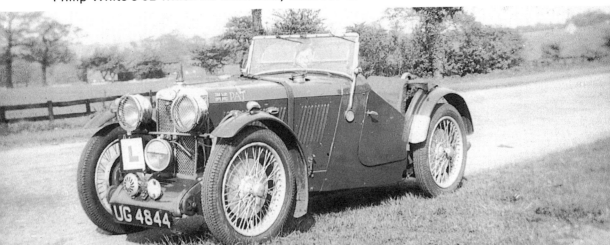

Bob Pattrick, the MG owner who first fostered the author's interest in the marque, with his Midget.

'Philip thought the car was a J4, but if you study the picture carefully you will see that the model has doors, which the J4 did not. He may have been misled by the big brake drums on his car, which were standard on the J4. I don't suppose Philip had ever seen a J4, of which only nine were made.'

Recalling this car which had fired his interest, Bob went on, 'When the J2 was introduced in 1932 it was greeted with rapture—a small, cheap sports car which was road-tested at over 80 mph. What the enthusiasts didn't know was that MG had slipped a souped-up model into the hands of the *Autocar*'s testers, and many a young blood—including Douglas Bader—broke their crankshafts in a futile attempt to emulate the *Autocar*. Production J2s would only achieve about 75 mph.'

He continued, 'I recall that every MG Midget after the J2 was greeted with scorn by the enthusiasts. The 'P' types, for instance, not only road-tested as slower than the J2, but had such effete features as trafficators! As for the 'T' types, with push-rod engines out of the Wolseley 10 saloon, instead of the high-tech overhead camshafts of the Js and Ps, they caused noses to be raised in disgust. More recently, the last, rubber-bumpered, Midgets have been regarded as inferior to the earlier, chrome-bumpered versions. All I can say is that my rubber-bumpered Midget is a civilised little sports car which is great fun to drive, and I think it fulfils the inventor Cecil Kimber's aims as well as any of the MGs ever made.'

This mixture of admiration and an awareness of the faults of the MG is, for me, one of the most remarkable elements of Octagon Fever. The car may

12

be loved and cherished, but no one would claim for a moment that it is perfect!

*   *   *

Until the summer of 1992 this book could have taken the form of a memorial to the MG sports car. Then, on a red letter day for enthusiasts, Rover released details of the first new sports car for thirteen years to bear the marque—the MGB RV8.

It had been back in 1979 that Sir Michael Edwardes, the head of British Leyland, had pronounced the fateful news that production of the famous little MG sports car was being halted. As workers at the Abingdon Factory in Oxford slowly ushered the last model off the production line, hordes of angry admirers of the marque, young and old alike, marched through London to protest at the demise of this legendary piece of British motoring history.

The actual end a few days earlier had been closer to farce than tragedy, according to *The Times*' motoring journalist, Kevin Eason.

'Senior BL executives, pressed into unfamiliar dinner suits, with their wives in spangly ballgowns, clinked glasses of champagne to celebrate 50 years of MG. Nearly 100 American dealers flew over for the party under the chandeliers of the Connaught Rooms in London's West End, hired for the occasion by the unusually spendthrift bosses at British Leyland. As the guests filed into the panelled dining room, the more cynical might have guessed there was to be a price to pay for such generosity. How high that price would be, they could not have known. British Leyland announced that MG was closing down. It was a bitter and shambolic end to a company which had become one of the most famous in the world.'

Sir Michael Edwardes was, naturally, seen as the villain of the piece, although he patiently explained that his mission at BL had been to rescue the ailing motor giant—and in the last year of its manufacture MG losses had amounted to £26 million. He said that the unique model was uneconomic, and the fact that the important American market for the cars was being decimated by the unfavourable dollar/pound exchange rate (it was calculated that in the summer of 1979 BL was losing something like £900 on every MG sold in the USA) made a shut-down inevitable.

Even Sir Michael, though, was amazed at the degree of public reaction to his decision. 'How *dare* the company get rid of a car that everyone loved and that evoked such strong nostalgia?' he was asked angrily. The protests came from literally thousands of devoted owners, not to mention some 70 Conservative backbenchers who signed a Commons motion opposing the closure. Writing later about those traumatic days in his memoirs, *Back from the Brink* (1983), Sir Michael admitted that his decision created more public fuss and misunderstanding than anything else in his five-year tenure at BL—even

13

The last MG ever made on the production line at Abingdon.

greater than the whole-scale factory closures and massive job losses.

'The public response and outcry was immediate,' he wrote. 'All our efforts to explain the commercial reality were nullified by this spontaneous outburst from outraged MG owners throughout Britain, and indeed throughout the world. Several hundred of them demonstrated outside the showrooms in Piccadilly and presented a petition to us. Questions were asked in the House of Commons and I suspect the whole tone of aggrieved outrage and anger which the demise of the MG generated prompted some of the opportunistic approaches to take over the MG business. One of them was from Alan Curtis, who at that time was running the Aston Martin Motor Company. I have no reason to suspect that his approach was other than a genuine and concerned attempt to save the MG name, but it certainly turned out to be a long-running and indeed highly damaging episode which we suspected from the start would come to naught and which in the event *did* come to naught.'

By the time Sir Michael wrote his memoirs, the MG badge had, of course,

14

been reinstated on a BL car—the MG Metro—which he claimed was being accepted by even the most fiercely loyal enthusiasts as a car in the MG tradition. But, he concluded, 'The moral of the episode is clear: you mess around with famous marque names that are loved and cherished by motor enthusiasts at your peril!'

The MG Metro and its successors—the Maestro and the Montego—were not sports cars, of course, but saloons, and could never truly replace the models that Cecil Kimber had first developed in the Twenties and which were sold by his company with increasing success for the next 50 years. However, the MGB RV8, unveiled at the British International Motor Show in October 1992, was beyond all question in that tradition, having evolved from the hugely popular MGB—although with a list price of £26,500 it was not the inexpensive sports car for every man which had been Kimber's dream and driving ambition.

The RV8 is, though, the first true two-seater sports car to be produced by any of the major British motor manufacturers for more than a decade, and boasts performance figures to match the best that Ferrari or Porsche can offer. With its 3.9-litre Range Rover engine capable of 0–60 mph in under six seconds and top speed of 135 mph, the RV8 is a soft-top for hard cash. 'Sixties nostalgia *is* becoming expensive,' one motoring journalist inspecting the first model was heard to remark, wryly.

If the 13-year hiatus had proved anything, though, it was that the MG was the sports car which refused to die. No new models may have been coming from the production line, but restored and renovated examples were

An epitaph on a late MGB roadster which has thankfully proved untrue! (The RoadRunner Collection)

MG owners 'Down Under' identify themselves by wearing very traditional English bowler hats! Bill Bennett of Victoria at the wheel of his 'P' type Midget. (Bill Bennett)

appearing everywhere in increasing numbers. It was estimated in the autumn of 1992 that at least 100,000 were still on the roads of Britain, with many more in Europe and America.

In the UK, two clubs, the MG Car Club in Abingdon (the older, having been founded in 1930) and the MG Owners' Club based at Swavesey, Cambridgeshire, were both flourishing. The Car Club can boast a worldwide membership of 40,000, while the Owners' Club, with a total in excess of 50,000, has been called 'The World's Largest One-Make Motor Club'. The Owners' Club has also instituted a campaign, 'Forever MG', in which owners pledge never to part with their pride and joy, 'subject to a personal disaster or other irreversible tragedy such as bankruptcy or death' and on so pledging are awarded a parchment certificate and special car badge!

On such evidence there can be no denying that the public's love affair with the two-seater sports car—and the MG Octagon in particular—is as strong as ever.

During some recent trips abroad I have also had the opportunity to see just how strong international interest is. On the other side of the world in Australia, for instance, I came across a whole bevy of beautifully maintained Midgets being rallied with every bit as much enthusiasm as I had seen in Britain—except that the owners of the pre-war models were identifying themselves by wearing bowler hats! Another enduring memory from 'Down

16

Under' was the sight of an MG special being driven in Victoria which bore the inscription, 'If God had meant racing cars to be rear-engined, MG would have made them that way'!

New Zealand has its MG enthusiasts, too, as do South Africa, Hong Kong, South America and even Japan. The largest number of clubs outside the UK is in Europe and, not surprisingly, the USA where enthusiasts rejoice in such names as the Abingdon Rough Riders (who are based in San Francisco), the Buffalo Octagon Association (from Brownsville, New York), The Emerald Necklace MG Car Club (centred on Cleveland, Ohio), Metrolina (in Charlotte, North Carolina) and The Uncertain T's (from—where else?—Oceano, California).

But Octagonal Fever is, after all, about individuals, especially people of character—one of the prerequisites, it seems to me, for any MG driver. As even the most die-hard enthusiast will admit, the cars often lack performance, are expensive to maintain, and are not that easy to drive. This is said to be particularly true of the MGB, the most popular model of all, yet still admirers invest many thousands of pounds each year to keep them running. As one owner confessed to me, 'Anyone will tell you an MGB is uncomfortable, cramped and noisy—and will keep you poor as long as you drive it. But the MG has a tradition that means something—it is a throwback that will look as good 30 years from now as it did 30 years ago when it was first built.'

Yet apart from such everyday owners of MGs (if they'll pardon the expression!) as my friend, there is also quite a number of familiar names bitten by the same bug. Indeed, the roll call of Octagon honour can be traced back as far as that daring aviator Amy Johnson, who was given an MG Mark I 18/80 saloon in 1930 to commemorate her 10,000-mile solo flight from England to Australia, through a group of Hollywood's leading film stars of the Forties and Fifties, to our present Royal Family.

The MG 'T' types—sometimes referred to as 'The Sports Car America Loved First'—were particularly popular with the film star fraternity, and no doubt the fact that people like Clark Gable, Gary Cooper, Nat 'King' Cole and Mel Tormé all drove TCs helped promote the marque among their fans. Mel Tormé was so infatuated with his TC that he actually wrote an article all about it for *Speed Age* magazine in May 1952.

'In late 1947, while doing my first stint as a solo singer at the Paramount Theatre in New York,' he wrote, 'I fell in love. The exalted object of my affection was a red MG TC Midget. Until then I happily drove a Buick Convertible and my interest in cars and racing was as a spectator. But the minute I laid my eyes on that wire-wheeled competent-looking piece of automobile, I was gone. I soon came to really appreciate the car, and later driving the little red job to and from the lot every day on the wide Los Angeles thoroughfares proved very rewarding.

'There was another side to the story, though, for I became accustomed to leaving restaurants and finding someone in the car "testing the wheel", while another stranger would be uncovering the motor "to see what the darn thing looks like". I'll always remember when Buddy Rich, the famous drummer and orchestra leader, caught a glimpse of me driving proudly down Broadway and stopped me. He begged a drive in the car round the block and then had me drive him to Zumbach's, where he placed an order for a black MG!'

The British actor Peter O'Toole is also an enthusiast and remembers with great affection a red 'T' type he drove all over Europe in 1958.

'I roared everywhere on the Continent in that little beauty,' he says, 'my bride-to-be, Sian Phillips, sitting alongside me. We went to Vienna and Berlin and all sorts of other places in it, looking for traces of Hitler, who fascinated me at the time.'

Another keen driver, Prince Philip, drove his black MG Midget rather more sedately when he started courting the then Princess Elizabeth in the closing years of World War Two. He did, however, make the most of the car's speed and manoeuvrability when motoring up from his naval base in Portsmouth to see the future Queen at Buckingham Palace. The Prince retained a real affection for the car long after his marriage, when jaunts in the MG became a thing of the past. We have, I am afraid, no information as to how often the Queen rode in that Midget—though we do know she enjoyed several outings in a TC owned by her uncle, Lord Louis Mountbatten, another Royal enthusiast of the marque.

Prince Charles has inherited his father's love of the car and owns a mineral blue MGC GT, number SGY 776F, which is currently housed in the Royal country retreat at Sandringham, Norfolk. The MGC was actually given to the Prince for his 21st birthday, when he became the Prince of Wales. Although the car has frequently been referred to as a Roadster, it is in fact a GT, and was specially prepared for its Royal owner at Abingdon before being handed over. The Prince later used the car for driving between London and Windsor, until it became too small for his requirements and was 'retired', initially to the Royal Mews in London, and then permanently to Sandringham, where it can now be seen by visitors.

Another MG with a famous owner that can be spotted in Norfolk is the blue 1972 MGB Roadster of Terry Waite, the Archbishop of Canterbury's special envoy, who was held captive in Beirut for 1,763 days. Since regaining his freedom, one of Terry's great joys has been to get behind the wheel of this roadster and head for the open roads of East Anglia, where he and his family now have a country retreat.

Among the other folk in the public eye who have Octagon Fever are Viscount Linley; the singer, Sting; TV personality, Angela Rippon; and the best-selling authoress, Charlotte Bingham. Charlotte recently explained the

Stirling Moss at the wheel of the MGB in which he finished third in the
1990 Pirelli Classic. (The RoadRunner Collection)

significance of the MG in her life.

'My brother offered me one of the first MGBs in 1963,' she said. 'I had
just been paid for my first book, *Coronet Among The Weeds*. It was a special
car with a superb engine and cost £850. The trouble was I couldn't drive,
so I took lessons from the instructor who had got all The Beatles through
their tests. He said I was far too young to pass and they would fail me if I
said I was an author. So I wore heavy-rimmed glasses, put my hair in a bun,
and said I worked for the Foreign Office. I passed first time!'

Another novelist with a fascination for the MG is John Wain, who actually
lives in Oxford. Indeed, the MG works at Abingdon and Cecil Kimber feature
prominently in his engrossing novel, *Where The Rivers Run* (1988). This is the
story of a working class boy studying at Oxford University in the Twenties,
who learns all about Kimber and his sports cars from an older brother, who
is employed at the MG factory. The following extract from the book gives a

flavour of the story in which the names of Kimber and MG run like a constant thread. In this passage, the narrator is asking his brother, Brian, just what is so wonderful about his employer.

'For once Brian really seemed to hear me. My words evidently stunned him. He put down his tea-cup and stared at me.

"You mean you've never . . . Well, stone the flaming crows. You've lived in Oxford all your life and you don't know who Cecil Kimber is?"

At this, my mother laughed: a sound rarely heard, for though she had plenty of humour, the responsibilities of her life tended to induce in her a mood of grave deliberation that seldom quite lifted. She often smiled, but rarely laughed.

"Brian, Brian," she said, "you are funny with your one-track mind."

"Thanks very much," he said.

"No, but honestly," she said, and laughed again. "You'd do better to turn it round and ask, 'How many people that have lived in Oxford all their life *do* know who Cecil Kimber is?'"

"Well, you do, for one."

"Only because you talk about him so much. D'you want some tea?"

"I'm not the only person that talks about Cecil Kimber," Brian said doggedly. "And before long there won't be anybody in Oxford that hasn't heard of him. And bloody few in England, if it comes to that."'

The famous American humourist S. J. Perelman was one of that country's best-known MG owners and regularly mentioned his 1949 MG 'T' type in articles, as well as several of his books. Sadly, though, he never recorded the story of his attempt in 1978 at the sprightly age of 74 to drive his Midget from Paris to Peking. Apparently, Perelman, who was certainly no novice at world trips, managed to obtain a sponsor for his marathon and had the car painstakingly tuned before preparing to leave the French capital.

However, even before he started, the author discovered that his female navigator had been signed up to write an account of the trip for a magazine—a fact which caused him promptly to dispense with her services. There then followed a series of arguments with his fellow drivers as they drove through India, compounded in Turkey by a row with Customs officials. On reaching Burma, the MG was refused an entry permit and, as there was no room for it on a ship bound for Malaysia, the car had to be air-freighted to Hong Kong instead. At this point, the other drivers decided to give up the whole idea, and then poor Perelman was struck down with a severe case of bronchitis, aggravated by double pneumonia. That was the end of the trip—and for a while the MG remained in a park at Kowloon until, like its owner, it could be shipped back to America where it has been preserved to this day.

Although Perelman did apparently try to salvage something from the disaster by starting to write an account for *New Yorker* magazine, this was still

Rallying and racing—two of the favourite pursuits of MG owners of both sexes, captured by the leading photographer of the marque, Ron Cover.

incomplete when he died just a few months later in 1979. The travel writer, Paul Theroux, one of the few people to have heard the outline details of his friend's epic MG trip, believes it would have been a classic piece of motoring journalism if Perelman had lived to finish it.

'Nothing is more Perelmanesque,' Theroux wrote shortly after the author's death, 'than a marathon drive across the world interladen with setbacks, blown gaskets, howling Turks and long delays in flea-ridden Indian hotels. And pneumonia in Peking was the perfect ending for someone who always racked his brain for grand finales.'

A verdict with which one can only, sadly, concur!

Perelman, of course, for a time worked in Hollywood, during which he became friendly with Barré Lyndon, the British screenwriter, whose fascination with motor sport—and the MG in particular—inspired him to write three classic books on the subject: *Combat: A Motor Racing History* (1933), *Circuit Dust* (1934) and *Grand Prix* (1935). Both men, no doubt, took pleasure from spotting their favourite cars whenever they appeared in the movies.

There have, of course, been fleeting glimpses of MGs in a great many films over the years, and it would obviously be impossible to list them all. Here are just a few of the more notable examples that have come to my attention.

One of the earliest pictures to feature the marque was *Car of Dreams*, made by MGM in 1932, which starred Greta Moisham, Robertson Hare and Mark Lester. It was highlighted by a sports car race at Brooklands, in which a number of MG Midgets competed with several Bentleys, Aston-Martins and Lagondas. These scenes were orchestrated by Rivers Fletcher, a young racing driver who was subsequently to become 'Motoring Consultant' for various British and American film companies.

They Flew Alone, made in 1943, in which Anna Neagle starred in the story of Amy Johnson's amazing life, also featured an MG. In several scenes the young actress was seen driving a 'J' saloon, ALW 522, through the streets of London.

The successful partnership of James Mason and Ava Gardner, in the movie *Pandora and the Flying Dutchman* (1951), was very nearly stolen by a superb MG Magnette driven by Rivers Fletcher; while Kenneth More had tremendous fun driving a 'P' type both on and off camera during the making of *Reach For The Sky* in 1965. More played the legendary Battle of Britain RAF Squadron Leader Douglas Bader who, despite the loss of his legs, was also an enthusiastic MG owner-driver.

Among more recent films, an MGB was extensively used by Michael Caine in the 1973 thriller classic, *Sleuth*, about deception and murder. Caine drove MGB LLU 755K in several scenes filmed at Athelhampton House, which doubled as the home of his co-star, Sir Laurence Olivier.

MGs have been featured on television, too, in series like *All Creatures Great and Small*, *Boon* and *Miss Marple*. One episode of *Boon* was notable for the appearance of a group of members of Hinckley MG Owners' Club complete with their cars for a scene at a Classic Car rally; while in the *Miss Marple* story 'The Body in the Library', one of the murder suspects, Basil Blake (played by Anthony Smee), was seen driving at speed through country lanes in a red Midget, ERB 240.

This year, 1993, the MG is at last going to become the star of a film, *Inside The Octagon*, made specially for television by the American film-maker, Gary Watson. Gary is mixing rare footage of early models in action with interviews with several of the people associated with the legend, including John Thornley, who was the post-war Managing Director of MG; Don Hayter, the designer of the MGB; Wilkie Wilkinson, a racing driver and mechanic; and, not surprisingly, Jean Cook. *Inside the Octagon* will have its première in America and, it is hoped, later be shown around the world.

* * *

The details I have outlined in the preceding pages point to one inescapable fact. Is it really so surprising that the MG—'the quintessential British sports car', as it has been called—should exert such enthusiasm among its admirers? Love of the car and its distinctive Octagon certainly grips people in all walks of life—and nowadays you are as likely to see an MG parked outside a modest, semi-detached house in a town estate as on the driveway of a large country house. It is an enthusiasm evident in just owning and driving an MG, or else through an all-consuming passion which demands the relentless tracking down of the tiniest parts so that a restored model is faithful to the very last grommet.

But is this fever just about nostalgia? Certainly, Cecil Kimber's original formula for an inexpensive two-seater is still just as relevant today, although the latest manifestation, the RV8, may be a little beyond the pocket of the average motorist. To my mind it is also symbolic of an enduring conviction that 'British is best'—or at least it should be. If only there was a home-made sports car that could compete with the Mazdas and Toyotas that currently fill our roads, how much happier, I have heard it said, we would be to buy it.

Perhaps, though, in a year or two it will be possible to satisfy this desire once again. For I have it on good authority that the next step in Rover's plans—if all goes well with the RV8—is to produce a new inexpensive, high-volume MG. The company is said to be already investing £100 million in developing 'two nippy, new convertibles'.

According to these stories, in 1994 the first all-new roadster in 32 years, code-named PR3, will be unveiled. It will have a transversely mounted mid-engine—like the successful Toyota MR2—and will look rather like 'a big

Midget'. It will be light and fast and powered by a 1.6-litre turbocharged version of the K-series four-cylinder engines used in the Metro.

Two years later, its big brother, a larger, front-engined two-plus-two, will hopefully materialise. Moulded along the lines of the Triumph Stag, this model will have a V6 engine developed out of the K series of between 2.5 litres to 3.0 litres.

Denis Chick, a Rover spokesman, while refusing to go into details of these two MGs-to-be, did admit earlier this year, 'We do want to go ahead with a high-volume MG—but we cannot base such a decision on nostalgia alone. We're moving back into MG because it's a sound business proposition, not just to make a bunch of MG traditionalists happy.'

Whether tradition alone motivates 'Octagon Fever' or not, the statement speaks volumes about the continuing appeal of the marque at the close of the twentieth century. And I have no doubt that the spirit of Cecil Kimber and his vision has much to offer the twenty-first century, too.

The Man Who Lived his Dream

Jean Cook

Jean Cook, the daughter of Cecil Kimber, the inventor of the MG, has been described as 'MG's best ambassador worldwide'. Indeed, her charm and her enthusiasm for the marque are so infectious that one motoring journalist recently referred to her as 'the biggest asset to the MG world'. Jean has painstakingly catalogued the Kimber family archives and owns literally hundreds of rare photographs of the early days of her father's career when the legend was just beginning. She remembers as a child feeling 'that something special was happening', especially when she was taken to visit the MG factory, given a peep at the secret prototypes, and even allowed to watch the racing cars being tested at Brooklands. Now living in retirement on the Isle of Wight with her husband, Dennis, Jean here shares some of her intimate memories of those early days of the MG and, in particular, of the man whose dream was to become a reality . . .

I—The Legacy

'Better to be born lucky than rich,' I thought, as I looked down from Rover's Visitor's Suite at the 1992 Motor Show, and watched the prototype MG RV8 revolving gently on its special stand. Tilted at a provocative angle, the gleaming torquoise body was surrounded by crowds five deep while my husband and I sat up above these eager admirers, in peaceful luxury, sipping our courtesy coffee. We were there thanks to our friend, Barry Price, the local MG agent on the Isle of Wight—and all because my father dreamed as a boy that one day he would build a sportscar.

Not all these admirers at Rover's MG stand were bona-fide MG owners; the kind of men and women who belonged to one of the many MG clubs, and could talk knowledgeably about all the various MG models produced by

Jean Cook with Stirling Moss at the launching of the new MG RV8 in 1992.

the little company my father had founded in the 1920s. Many of them, indeed, readily admitted they had no hope of owning even a clapped-out secondhand MGB, but they just 'loved' MGs and all they stood for.

Well, what do they stand for? Why are they loved so passionately by their owners, and by people who can never hope to own one? My husband and I toured the world in 1988, the year of my father's centenary, celebrating with enthusiastic members of MG car clubs in South Africa, Australia and the United States, and everywhere I asked, 'Why MG? Why don't you treasure Fords or Austins or Rileys?' I got various answers. 'It's such a good design', one owner said. 'There are so many interesting models,' added another. But the one I liked best, and it is not so fanciful as you might think, came from Bill Bennett, a tough Australian owner of a pre-World War Two 'P' type Midget, which he had spent two years of hard grind restoring, who said, 'Because your Dad put a little bit of his personality in all his cars.'

The new MG, of course, has sprung from the MGB, the best-selling sports car of the 1960s and '70s—the brainchild of Sid Enever, Don Hayter and the design team at the MG Car Company, Abingdon-on-Thames, presided over by John Thornley, the Managing Director, who started out as a young man in the service department in the early '30s as a direct result of helping to found, and steer as the first secretary, the MG Car Club. He told my father he couldn't go on running it, unless he had a compatible job!

The crowds surrounding the new model had mostly never heard of Cecil Kimber, who was killed in a train crash in 1945. Indeed, the young executives on the Rover stand hadn't heard of him either—to the chagrin of Mark Brotherton, our host, from British Motor Heritage, where the idea of a reborn MG first stirred. But I was not surprised. After all, my father died before they were born, and in Mark's case, probably when his father was still at school. Inevitably, though, I was asked what my father would have thought. My stepsister, Bobbie Walkinton, always says that he would have expected it. She means, of course, both the passion for restoring and even racing all the early models, and then the post-war development of MGAs, Bs and Cs pioneered by the men who started at MG when he was in charge. I can

Jean Cook, aged 15, with her father at Bridge Cottage in Gloucestershire in 1940. The photograph was taken by Gillie Kimber.

imagine him saying, 'Well, of course they'd go on making the best car in the world.' Of course it *isn't* the best, but it seems to make the most friends.

I had the good luck to meet Stirling Moss only three days after our Motor Show visit, when my husband and I were guests at the official opening of the MG Owners' Club's new HQ at Swavesey, just outside Cambridge. Stirling and I spent quite a time sitting in the MG RV8 (which had arrived on a transporter) discussing his view of the new car, which he feels is a fast, comfortable tourer, but not really a true sportscar that an enthusiast could strip down and take racing at Silverstone. However, we learned that Rovers did some very careful research before they launched their new MG. They envisaged a well-off family with two or three cars already, intrigued by the legend and pedigree of MG, but demanding the performance and reliability of their modern cars. I pointed out to Stirling that Dad had a lame leg which hurt him in cold, wet weather, and that he adored his own fast touring saloons, which Lord Nuffield felt would be more profitable than the way-out racing 'R' type of 1935. I said I thought my father would have enjoyed taking this car to the Continent, on his annual visits to MG agents in Switzerland and Germany, and Stirling agreed that he himself would not mind that chore at all! If 135 mph as a top speed seems rather excessive in Britain, it is not on Dutch or German motorways which are, of course, de-regulated.

Stirling was there by virtue of his genius as a racing driver. He was invited to pilot the new MG Ex 181 in 1957 to try—on the temperamental Bonneville Salt Flats in Utah—for the Class F (1500 cc) record which MGs had held since Goldie Gardner achieved 204.2 mph on one of the new 'autobahns' in Germany just before the outbreak of war in 1939. On the evening of 23 August, after waiting patiently for conditions to be just right, Stirling broke five Class F records with an average two-way speed of 245.64 mph.

I found Stirling a delightful character to meet and to listen to as he unveiled the commemorative plaque in the entrance hall with a chequered racing flag, and he convulsed his audience by heaving a sigh of relief that his name had been spelled correctly. He was always afraid he would find it permanently bronzed as 'Sterling'—'not a welcome allusion at the present time', he said, amidst laughter that had a touch of apprehension mixed up with the mirth.

Opposite the plaque, there hangs a framed print of the study of my father in oils by Robert Anderson, an American MG enthusiast as well as a portrait painter by profession. He created the original from photographs I lent him for the MG 'T' Register of New England, an enormous organisation devoted to driving, restoring, rallying, racing and just enjoying the various 'T' type models about which you will find much more detail later on. I saw the original portrait and met Robert when we were in America in 1988 as guests of the Chairman of the 'T' Register, Dick Knudson, who organised the Kimber Centenary celebrations, including the publication of *The Cecil Kimber Book*. It was at this time, too, that I gradually began to solve the puzzles presented by so many unidentified photographs I possessed, and fill the gaps in the story of how he came to found a little car company that has given so many people in so many countries, over so many years, so much fun and adventure.

Jean Cook with 'Old Number One', the car mistakenly believed to be her father's first model—though it does still run today!

II—Beginnings

Cecil Kimber was born on 12 April, 1888, at West Norwood. He arrived in the world about the same time as the internal combustion engine, and may well have seen the first London to Brighton Run in 1892, but he rarely thought or talked about the past. His sights were set on the future—mostly on what the next MG model would be like. But from odd remarks, and stories told by our mother, we—my older sister, Betty, and I—grew up with the impres-

29

sion that my father had suffered a miserable childhood. His father, Henry Kimber, always referred to as 'HK', was tyrannical, gloomy, demanded that children should be seen and not heard, and, most heinous crime of all, would not buy him any rails for his model steam train. Most writers have copied this version from Wilson McComb, who talked to us in the 1960s. It is of course true that Cecil Kimber was apprenticed to his father, a manufacturer of, not glamorous cars, but printing ink. How dreary, we thought!

After the centenary celebrations I had a chance to visit a second cousin, Marie Kimber, in New England. Her son, Sidney, is the image of my father, even down to the habit of tapping his knee when talking! I learned from an account of a visit to England by Sidney's grandfather (my father's favourite uncle) in 1923, that Henry, my grandfather, could not help his gloom and his longing for quiet. As an adventurous young man, he had marketed, with his brother-in-law, ingenious, wheel-mounted luggage carriers for penny-farthing bicycles, but careering along the rough tracks of the New Forest in the 1860s, he crashed. There were no helmets in those days to save him from concussion and a lifetime of blinding headaches. By luck, or good sense, he chose as a wife Fanny Matthewman, a beautiful and talented artist from Huddersfield, who also needed peace and quiet to produce her lively, impressionistic landscapes in watercolour that were exhibited in her lifetime, and are still listed as 'collectable' by Bradford Library. Her father, Sidney Matthewman, worked for United Alkalis, a forerunner of ICI, and was an agent for early car batteries.

I have also learnt not to despise the printing ink. HK was descended from a father and grandfather who were pioneers in printing engineering in the early nineteenth century. Richard Hughes invented the steel plate for printing art work; his son-in-law, Edward Kimber, introduced the first machines to print from lithographic stone by steam in Britain and America. They both became Freemen of the City of London, and Edward made another fortune from introducing lithoprinting into China, for which he had a period monopoly. His seven sons inherited not money, but shares in the family business, which began to decline under the less-than-brilliant management of Richard, the eldest son. HK decided to join two of his older brothers, Walter and Edward, who had moved to Manchester and founded Kimber Bros, a printing engineering business, but they were all forced, by their father's will, to leave their capital in Hughes and Kimber. Walter and Edward died, leaving HK on his own, and as he had managed the Hughes and Kimber varnish factory in Mitcham, it is logical to guess that he knew how to manufacture most of the subsidiaries for printing. So the young Cecil grew up with a tradition of invention, enterprise, and making a new start. He did not want to be a printer, or even a printing engineer, but he knew how to organise a small factory. This was to become his major talent.

30

After the migration to Manchester, Henry and Fanny Kimber, with their sensibly-sized family, Cecil, Phyllis and Vernon, settled at Heaton Moor, in a large house called 'Moorfield'. They had a maid called Mary, and seaside holidays—all photographed by the young Cecil, who stored and labelled all the negatives in one of his exercise books from Stockport Grammar School. He left with a School Certificate but had no hope of training in mechanical engineering, though he did attend evening classes at Manchester Technical School 'taking quite the wrong subjects'. Speaking to graduates of the Derby branch of the Institute of Automobile Engineers in October 1944, he said: 'All I had was an overpowering attraction towards motor cycles and motor cars in general. Compared with many of you here, my early engineering training was of the most sketchy nature. It must be remembered that in the early 1900s the opportunities for technical training in automobile engineering were very scant as compared with . . . today.' He went on to say that the only subject of any use at Technical School was accountancy . . . 'after all, you go into business to make money. If you don't make money, you won't remain in business very long, anyway.'

The Kimber family by this time were living in a charming old house called Greenbank, in the Cheshire village of Grappenhall, which became the home my father always remembered. He had a bicycle, and the freedom to ride 14 miles to Dunham Hill near Altrincham 'to watch the very few motor cycles and cars coming back to Manchester from their Sunday afternoon run . . . this modest incline used to test their cooling systems to the utmost.' For his next adventure, which was to alter the course of his life, he had to save up his money.

He bought his first motor cycle in 1906 when he was 18—a secondhand $3\frac{3}{8}$ hp Rex, price £18, which he soon stripped down to find out how it worked. From an article he wrote for *The Motor Cycle* in 1908, we know that he gradually made longer and longer runs with the Warrington and District Motorcycle Club, and he learned from a Mr Wolstoneholme, and from his friend Oscar Whittle of Whittle Belts, how to save money by making his own repairs. It was a short step to racing against motor cars—unofficially, with other WDMC members, on their runs back from Wales. In 1909, he bought a 1907 Twin Rex, and after a night spent opening the ports, he won his first unofficial race held at dawn, against a member who owned a 1909 Triumph. He was riding his future brother-in-law's new Rex in 1910 when he was hit by an elderly solicitor in a car at Grappenhall crossroads. Ironically, my father was going quite slowly—his mother had invented an errand for him to get him out of the hated chore of mowing the lawn.

He smashed his right knee and thigh and spent the next two years on crutches, in and out of hospital while the surgeons tried to save his leg. Eventually, as they were thinking of amputating, the bones began to knit.

31

The motorcycle accident in 1910 which was to have such a profound effect on Cecil Kimber's life.

He could walk, he could even dance and learn to skate, but above all he could drive. What did it matter if he walked with a limp, and his leg ached in cold, wet weather? This was the price of immunity from the carnage of the First World War, and the compensation awarded him, said to be £700. He spent some of it on an early model Singer 10, which he used for selling the printing ink his father produced, and in 1913 he wrote an article for *The Light Car* to show how car-ownership had saved time, cut costs and increased his sales.

During the two years that my father was on crutches, he also had to face the death from cancer of his beloved mother. She is buried in Grappenhall churchyard and, though of course I never knew her, I feel very close to her as I have her paintbox and a brush most cunningly clipped for painting trees, and several of her delightful landscapes. My favourite, painted at Abererch in North Wales, has trees you could swear were swaying in the wind. Thirty years later, her son was to demand that his artist paint his cars 'to look fast standing still'.

In a lecture in 1945, my father said he learned to drive when he was quite young, in Wales. His parents first spent their holidays at Runswick Bay, a famous gathering place for artists in the 1890s on the north Yorkshire coast (where my father learned to row and sail from a local lifeboat hero, Ned Clark). Then they switched to Penclogwyn Cottage in Abererch, near Pwllheli. I suspect this was a Hughes family inheritance in a tin-plating area. When we visited, a local historian, Mr Owen, told us my father used to play with Hugh John Williams, a future canon of Wrexham Cathedral. I wonder if Hugh's father owned the 10 hp Wolseley 'with the gear change on the steering wheel' in which my father remembered driving 'in reverse and in the rain from Beddgelert to the Pen-y-Grwd Hotel, to see the competitors in the Six Day Trial attack the Llanberis Pass'?

We know from photographs that a friend called Gilbert brought a T-head Singer to Abererch which Bill Boddy, founder of *Motor Sport*, has identified as the racing car owned by Vivien Hewitt, the first pilot to cross the Irish Sea. My father bought the Singer later, after it had acquired a Hispano-Suiza body, and said, in one of his last talks in 1944, that it was rare in having a fourth gear and was reputed to have lapped Brooklands at 80 mph. Hoodless, springless, comfortless, he took my courageous mother for a honeymoon tour in it, and showed her Penclogwyn Cottage. He photographed the cottage, and then the empty lane. Now why?—unless it was there he first had the vision that one day he would build a racing car which could win at Brooklands.

III—MG is born

My mother, Renee Hunt, was 'one of the belles of Withington', according to her youngest sister, 'and could have married anyone.' She was the eldest of six forthright, well-educated sisters, and worked as a translator for a Manchester cotton firm. Her father, Charles Hunt, was an engineer with Selson Ltd, a manufacturer of agricultural machinery, and her mother had been a teacher at Leeds High School for Girls. They lived in Fallowfield in a typical tall, thin, red brick house, which resounded to my Uncle Vern playing the piano, and a crowd of young people singing the latest hit tunes. But what did the beautiful, intelligent Renee see in a small, lame salesman of printing ink earning £1 a week? I think she saw adventure.

All his life, my father had the ability to make even the simplest picnic into a memorable event. His enthusiasm was infectious. He took my mother out and about—a novelty for a girl brought up by a father who did not dream of including his family in his outdoor pursuits. By this time, my father had the re-bodied Singer and they went to hill climbs and rallies run by the Manchester Motor Club and the Lancashire Automobile Club. When he asked Renee to marry him, she agreed. Renee's father warned him about her temper and their inner lack of compatibility, but my father would not listen.

HK, now widowed and living in Withington, Manchester, far from giving his son a rise on which to get married, wanted him to hand over the balance of his compensation money to prop up the printing ink business. My father refused. They had a furious row, and old HK retired to a cottage near Towyn and never spoke to his unfaithful son again—though my father tried many times to heal the breach. My mother encouraged the rebellion, and by 1915 my father was personal assistant to the chief engineer of Sheffield Simplex, and living with his young wife at Norton Woodseats. Surprisingly, he rode a Sun Villiers motorcycle to work. You would think he would never have wanted to touch another motor cycle, but in fact he won medals for hill climbs and speed trials when he first came to Oxford.

From Sheffield Simplex, my father moved to AC Cars at Thames Ditton, where my mother was his secretary, and then to Martinsyde Aircraft at Weybridge as a stores organiser. In 1919, he went to E. H. Wrigley in Birmingham, as works organiser and, according to my uncle, 'you could have eaten your dinner under the machines when he had finished with the place.' He was said to have designed the radiator for the unsuccessful Angus Sanderson mass-produced car, in which Wrigleys had unwisely invested. In a lecture in 1944, he said he lost his savings in this disastrous venture, pointing out the lack of wisdom of respected manufacturers 'having all their eggs in one basket'. William Morris (later Lord Nuffield) the astute founder of the Morris Motors empire, who bought Wrigleys, spotted my father's gifts and offered him a job as sales manager at Morris Garages, the retail sales and service business, quite separate from the Morris main plant at Cowley. In two years my father was general manager, and beginning to design special bodies on a Morris chassis—the first MGs.

The famous octagon badge was the brainchild of the Morris Garages Oxford accountant, Ted Lee—but there have always been arguments about what MG stands for, and whether the letters should have full stops after them. In an advertisement in the late '30s, my father stated quite clearly: 'So many people who do not know, ask: "What do the letters M.G. stand for?" They don't stand for "More Ginger", though that would be applicable. They were given as a compliment to Lord Nuffield, being the initial letters of Morris Garages, which was his original business, and from which the M.G. in particular and all his other vast enterprises have sprung.'

However, when I was a child, I remember quite clearly my father saying, 'MG doesn't stand for Morris Garages, it just stands for itself.' But when I produced this idea in my speech to a 'gathering of the faithful' at Killington, Vermont, in 1980, the assembly was profoundly shocked. I had sinned against Holy Writ! But I stuck to my guns and in 1991, Robin Barraclough, secretary of the Bullnose Morris Society, after extensive research amongst Lord Nuffield's papers in Oxford Library, announced that Lord Nuffield's accountant had insisted that the infant M.G. Car Co must be separated entirely from Morris Garages or his noble client would be liable for supertax. Thus my father insisted that M.G. just stood for itself. When I mentioned this in my column in *M.G. Enthusiast*, Jonathan Daniel wrote from Northumberland to support me. He had heard my father, at an RAC awards dinner, announce that he had chosen the M from Morris and the G from Garages 'as a tribute to his employer' and that M.G. certainly did not stand for Morris Garages. There was already a Morris Garages, so how could there be two? 'He made it clear to all present . . . M.G. stood just for itself—The British Sports Car'. When queried about the full stops, my father said, 'They looked more visually attractive.' I still like MG on its own!

Not many people know that my mother played an important part in the development of the first MGs, and actually helped with the early designs. She charmed William Morris into supporting the new venture, and, more essentially, also persuaded H. N. Charles, the Morris Motors designer, to help 'the Kimber team' in the evenings and at weekends. He was to tell Wilson McComb, the MG historian, that 'she was a most cultured, charming, wonderful person.' Sadly, she never really recovered from my birth and died, after a long and debilitating illness which I now suspect may have been a rare form of cancer. But she lived to see the descendants of the early MGs winning at Brooklands and on the famous continental race tracks, and she had several Midgets, including a very lively J2 in the saloon version, with which she competed in the early MG Car Club rallies. She was a marvellous hostess and a loving mother until illness and despair began to conquer her natural charm.

We lived in Oxford, first at 339 Woodstock Road, and then in a large Victorian house, 1 Hernes Road, where my father had electric trains running round our nursery. Lord March, heir to the Duke of Richmond and Gordon, who was to promote motor racing at Goodwood, loved coming to play with them, and even helped my father get his model steam train going at long last. My grandfather, you may remember, refused to buy rails for it. I am not surprised: it was a terrifying little bomb when it got steam up.

The design and building of a whole range of MG Midgets, Magnets and Magnettes in his new factory at Abingdon during the early '30s must have fulfilled my father's dream. As an impressionable 7–10 year old, I watched some of the supercharged versions in the great races of that time at Brooklands, and even had the luck to be left in the racing pit one day by mistake. I sat unnoticed, watching enthralled as the brilliant racing mechanics, under Reg Jackson, changed tyres with lightning speed and refuelled using ordinary petrol cans and a huge, home-made funnel, while the luckless driver was casually covered with a tarpaulin. I was too young to know that the team had probably been up all night tuning the cars and sorting out problems discovered during practice. I simply thought they were heroes—and the drivers were young gods.

My father honestly believed that success in racing, particularly on the Continent, helped a small manufacturer to become known worldwide, particularly a company such as MG, whose racing cars could be bought in sports versions by the ordinary motorist. In an article, he pointed out that 'very many lessons learnt in the stress of racing were embodied in the subsequent production cars, to the ultimate benefit of the user. Brakes, steering and engine-bearing life were perhaps principally affected and improved.' He tended to ignore the problems faced by these ordinary purchasers when their dream car turned out not to be as fast as the supercharged racers they saw scalding round Brooklands. His own supercharged Corsica-bodied K1 Magnette, specially built for him, and which was our family transport for two years, tended to be a headache for the service department because engine smells would leak into the passenger area. I didn't care about that—one magical, never-to-be-forgotten day I saw the speedometer top 100 mph on a quiet, straight road in the Cotswolds!

(Below) Boundary House, the Kimber family home in 1933 which is now a public house; and *(opposite)* a family group at the house in the same year, including a shy Jean in front of her father and, beside him, her mother, Renee Kimber.

By this time we had moved to Boundary House, Abingdon, about a mile from the factory, where my father could bring customers home for lunch, often without warning. But our wonderful cook, Winifred Carter, aged 18, coped with everything, while her brother, Harry, kept the large garden in order. We had dogs, cats, hens, a pond, an orchard, a tennis court, a tiny wood for my wigwam, a larger outdoor playroom for the trains, and a hutchful of piebald mice.

Absorbed in the excitement of developing the more advanced racing machines of 1934/5, Cecil Kimber perhaps assumed too easily that Lord Nuffield would back him, as he always had.

I cannot help wondering if my mother encouraged him fatally. Why else should he say long afterwards that her interest was welcome, but not when it became 'a whip to drive me'. The only clue I have is that my mother once took me with her when she went to talk to H. N. Charles, the MG chief designer, when we lived in Oxford. We walked round to his house, and I played with his roulette wheel—a fashionable pastime at parties then— during the discussion. By the tone, it was clearly 'business' talk, not emotional. My mother could no longer play any role in MG design, and she was keeping up with developments through old friends.

But it was too late. The 'R' type, though full of innovative design which would be exploited in the Morris Minor front suspension after the war, was too much an out-and-out racing car. There were expensive teething troubles and this time Lord Nuffield looked at the balance sheet and said 'no'. There must be no more racing. MGs must build 'family' saloons alongside the new 'T' type Midgets. The MG Car Co was sold to Morris Motors, my father lost his independence and found himself responsible to Leonard Lord, the abrasive cost-cutter from Birmingham, who was made Managing Director, while father was demoted to general manager, though also a director of Morris Motors.

I find it interesting that he was not more upset by this major upheaval. He actually liked Len Lord, and he welcomed the challenge of the saloons and indeed a 1½-litre (now called the VA) became his favourite car. The warmth and comfort of a fast saloon was not lost on him. He was, after all, turned 50, and had that painful lame leg. He was responsible for the beautiful lines of these cars—Winifred, now our housekeeper since my mother was in a nursing home, remembers him bringing the designs home to work on until he was satisfied. At this time he read a paper to the Design and Industries Association—'The Trend of Aesthetic Design in Motor Cars'—in which he refers to this work. He went fishing in Scotland in his new VA; and he went back to his first love, sailing, and bought a cruising yacht, which he kept on the Beaulieu River. He had always loved photography and now he bought a Leica.

Renee Kimber, with the Earl and the Countess of March, and her 'M' type Sportsman's Coupé.

The end of the MG Works racing department did not mean the end of preparing private owners' cars, nor of rallying, which became very popular. My father obtained permission to build a special record car for Major 'Goldie' Gardner, a friend of long standing, and had the great satisfaction of watching the Railton-bodied special, nursed by Reg Jackson and an experienced team from the works, and driven brilliantly by Goldie, attain 210 mph on a German autobahn.

My mother died in the spring of 1938. She had been separated from my father for the past year and lived with my sister Betty (and me, during school holidays) in a sunny flat overlooking Poole harbour, a fact which had not escaped Lord Nuffield. However, now my father was at last free to marry the gentle, witty, wholly supportive divorcée, Muriel Dewar, nicknamed 'Gillie', whom he had loved since 1933. I am sure that Gillie helped him through 'the bad time'. They had felt that they should wait until my sister and I, and our stepsister, were grown up, but the Munich crisis pushed them into a secret marriage earlier than they had intended. When the crisis resulted in an uneasy 'peace' for a year, they moved into the Miller's House, Pangbourne, a tranquil paradise with its own trout fishing. Having dreaded the thought of a stepmother, I loved Gillie on sight, and continued to love her until she died in 1980. She transformed my father's life, which now became full of jokes, teasing and hilarious parties which blossomed out of solemn business dinners. To this day, I treasure the memory of a rather pompous guest who ended the evening crawling under the table, almost weeping with laughter,

Photograph by Cecil Kimber of his specially-built family car, the supercharged Corsica-bodied K1 Magnette, in which Jean Kimber travelled at 100 mph.

to find the cherries he had dropped. Gillie had taught us all to 'bob', i.e. to nibble up from the end of the stalk and try and hook the fruit into our mouths.

The Second World War smashed this idyllic existence in 1939. My stepsister joined the WAAF; Betty, now Mrs Delamont, was designing camouflage for aircraft factories; and I was still at boarding school, though I joined the WRNS later. My parents stayed on at Pangbourne until they moved in 1941.

The outbreak of war left MG with no work. This seems odd when Morris Motors, then under the direction of vice-chairman Oliver Boden, was overwhelmed, both at Cowley and their new aircraft factory in Birmingham. Lord Nuffield was seconded to the Air Ministry, as Director-General of Maintenance and, in the confusion, MG seemed to be forgotten. My father and his works manager, George Propert, went out and found their own war contracts. By 1940, they were repairing tanks for General Peto, Regional Controller for the Ministry of Supply. This came through a fishing friend, John Howlett, Managing Director of Wellworthy Piston Rings, who had been appointed chairman of the Emergency Services Organisation, Southern Region, in 1940 under Lord Beaverbrook, who knew that all our war production would be bombed. Howlett set up, with my father's help, a secret plant to make pistons for Spitfires in part of the Pavlova leather factory, next door to MG. He also leased Fyfield Manor, only five miles from Abingdon, as his HQ, and my parents lived there, with my stepmother's sister, Gladys Hamilton, liaison secretary for Wellworthy's, who organised war production meetings there. She later became the second Mrs Howlett.

Sailing was the other great passion in Cecil Kimber's life. *(Above)* With Jean at Buckler's Hard in 1937. *(Below)* Cecil's second wife, Gillie, sailing on the East Coast in 1937–8.

By this time, Miles Thomas, a 'corporation man' to his fingertips, had succeeded Oliver Boden, who died suddenly of overwork in March, 1940. In his autobiography, Thomas claims my father was far too independent: 'privately, Morris had never liked the idea of sponsoring a sports model . . . but Kimber persisted. He was a brave man, took periodic wiggings for lack of profit-making with impudent sang-froid, confident that prestige gained in sporting events by the MG car one year would be recouped in sales the following year. He was an originator; he designed (and) had everything possible made octagonal . . . This kind of philosophy was all right during peace time, but when Kimber wanted to maintain his acute individualism after the war had broken out . . . it was clear there must be a change.'

Because of the connection with Lord Beaverbrook and John Howlett, Thomas could not quite use MG's next decision to tackle the complicated

building of the 'noses' of Albermarle bombers as an excuse for getting rid of a thorn in his flesh, so he bided his time. According to my father's story, a directive came from Morris Motors to centralise the issuing of unemployment and insurance stamps, which would mean sacking a faithful MG employee, a single woman who supported a widowed mother. My father refused. Next day, Miles Thomas arrived to demand his resignation, and wrote later: 'He was thunderstruck, but took it with brave grace . . . I was sorry for Kimber. He was completely unorthodox even in his domestic life, which did not endear him to Lord Nuffield, nor, in particular, to Lady Nuffield.'

Many MG writers have expressed dismay at this needlessly brutal dismissal at such a time, when our aircraft losses were catastrophic and we were losing the war. There would appear to have been an old score to settle rather than any fault on my father's part. He accepted his fate with quiet dignity, called his workforce together and said goodbye without any explanations, and scotched offers to strike in protest 'as there was a war on'. He drove home to Fyfield Manor, white and shaken, to tell John Howlett what had happened. 'Go and see Nuffield,' said John, but Lord Nuffield talked platitudes and would not help. 'Sue them,' said John, but my father said there was a war on, and went quietly away, first to his friend Charles Reynolds, to reorganise Charlesworth Bodies at Gloucester, and then to a much better paid, but far less satisfying, job as works director of Specialloid Pistons. There, he and the technical director, Eric Graham, fitted Eric's invention—an extra metal collar—on to tank pistons, which gave British tanks 400 extra miles before they required servicing and proved a crucial factor in the battle of El Alamein in 1942.

By the end of 1944, my father was giving lectures on the future of sports car production, appearing on panels, and planning to live in a cottage at Itchenor on Chichester Harbour, doing some consultancy, and sailing the beautiful ketch he bought in 1944. He survived the V1s and V2s which showered on London in the summer of 1944, and then, ironically, in February 1945, just a few months before the Allied victory, he was killed in a stupid accident in a train reversing at walking pace in King's Cross station. He was 56. 'So when Nuffield turned round at the end of the War,' wrote John Howlett, 'when the motor industry was in a ferment of innovation, and said, "What we need now is a Cecil Kimber", nobody could provide him with one.'

I don't think I have ever got over losing him before I really knew him, as an adult. My father was an extraordinary man—energetic, enthusiastic, sometimes quick-tempered for the wrong reasons, sometimes obstinate, or blindly pig-headed—but he has left us all an incredible legacy, a whole world full of friendship and fun, the world of MG—or perhaps, this once, I should write M.G!

42

The Vintage Years

Cyril Mellor

Turning Cecil Kimber's dream into reality was to demand determination and hard work from the dedicated team that the inventor assembled in Oxford. The details of this crucial opening chapter in the MG story are related here by Cyril Mellor, the secretary of one of the clubs associated with the oldest models, The Vintage Register. Cyril is a particularly appropriate person to be writing this piece of history, for he has also been responsible for restoring one of the very earliest MGs, a Mark IV, believed to be one of only four still in existence. His account of how he painstakingly rebuilt the car from its almost completely dilapidated state to pristine glory, not only brings history alive, but underlines the magic that is inherent in the MG name, whatever the vintage . . .

To find the beginning of this unique story in British motor manufacturing history, and indeed the birth of a whole sporting activity, we have to go back to the very earliest days in the career of William Morris, later Sir William Morris and, eventually, Lord Nuffield—a car maker who became the philanthropist whom we have to thank for the start of the Nuffield Hospitals.

William Morris was a visionary rather than an inventor. He could visualise how to bring together well-engineered mechanical components from a variety of manufacturers in order to produce a reliable motor car at a marketable price.

His early experience as a repairer of cycles, motor cycles and cars had taught him the weaknesses and strengths of the many types of automobiles then produced.

The early Morris Oxfords, powered by a White & Poppe engine, followed by his Morris Cowley, fitted with an imported American Continental engine, eventually to be built under licence in the UK by Hotchkiss of Coventry,

Cyril Mellor with his wife and the 14/40 he restored from scratch,
photographed during their tour of New England in 1991.

were prime examples of his buying-in policy. Other early main suppliers
were Sankey, Alford & Adler, E. H. Wrigley and Raworth of Oxford for car
bodywork.

This method of producing cars without the need to purchase expensive
machine tools continued after the First World War with increasing success.

Many other competing car manufacturers of that period found themselves
with excellent and well-equipped factories but with high overheads, that could
not be supported by the low volume production of their specialist cars.

In the boom and bust period of the early 1920s, many of these specialist
car firms were forced to close but, because of his buying-in policy and lower
overheads, William Morris was able not only to survive but thrive.

Before Morris started making his cars, his repair business in Longwall
Street, Oxford, was known as Morris Garages. When car production commenced in 1913, this location became his customer service depot, under a
separate works manager.

Cecil Kimber, another man with a vision, and the drive to make it a reality,
had by this time already enjoyed seven years' experience in the motor car
and component manufacturing industry with the firms of Sheffield Simplex,

A. C. Cars, Angus Sanderson, and E. H. Wrigley. In 1921, at the age of 33, Cecil Kimber was appointed sales manager for Morris Garages and within a year became general manager.

It was during the early Twenties, as it became obvious that the motor car was rapidly replacing the horse carriage, that the sporting gent had a desire for a car in which to express his personality. The heyday of the affordable sports car was about to commence.

Of the surprisingly large number of British car manufacturers in the last 90 years, no fewer than 140 have listed at least one sports car in their range. Head and shoulders above all these producers stands MG, whose Octagon badge has adorned more cars than all the other British sports cars put together.

The man behind the design of the whole of the MG car concept, from its first beginnings in the early Twenties through its sporting, racing and record-breaking days in the Thirties, was Cecil Kimber. His design influence endured until the last of the traditionally built 'TF' series cars rolled off the Abingdon assembly lines in 1955.

His first efforts resulted in six Morris Cowley chassis being fitted with special open two-seater bodies built by Raworth, the Oxford coach builders. Sadly, none of these original six cars seems to have survived, although copies of an advertisement for 'The MG Super Morris', with Mrs Renee Kimber at the wheel, fortunately have. They show the car with its lowered suspension, scuttle ventilators and unusual five-piece windscreen appearing distinctly

A rare photo of one of the six original MG Super Sports Morris cars with open two-seater Raworth bodies. (Early MG Society)

sporting when compared to the Morris Cowley two-seater tourer of that period.

Not only did the Kimber MG look sporting, its mildly tuned engine, higher rear axle ratio and lighter overall weight gave it an appeal and performance that enabled it to be marketed at a price considerably higher than its Cowley counterpart.

Although the special sporting two-seater MG powered by an overhead valve Hotchkiss engine (of the type produced for the Scottish-built Gilchrist cars), in which Kimber gained a Gold Medal in the London to Land's End Trial of 1925, is usually referred to as 'Old Number One', around 50 special bodies had been produced by Raworth for the MG Super Sports Morris up to this time.

Certainly the advertisements for the Raworth-bodied cars of 1924 bore the MG Octagon, and cars produced in early 1925 carried the logo on the door sill plates. Nevertheless, the car every MG enthusiast has come to refer to as 'Old Number One' embodies all the sporting characteristics we have come to recognize from the most successful production sports car designer ever.

Its chassis frame was basically of the type used on the 1924 Bullnose Cowley, but with the rear end shortened and modified from its usual over-and-under three-quarter elliptical springing, to almost reverse camber half-elliptical springing with the rear axle bolted beneath them. This lowered rear suspension was fitted with Hartford adjustable friction shock absorbers, whilst the flattened spring front suspension was equipped with Gabriel Snubbers, a device that only dampened the spring rebound action.

Early photographs taken during the Land's End trial clearly show the car equipped with Morris Oxford-type 12-inch front brakes. At some later stage these have been substituted by nine-inch Cowley type. Surprisingly, the car was equipped with the narrow bead edge tyres, since Morris had already introduced balloon tyres in late 1924, whilst for the first time, probably in order to obtain better weight distribution, Kimber used a rear-mounted petrol tank.

For some seven years following the sale of the car by Kimber to a friend in Lancashire, virtually nothing is known of its history, until it was discovered in a scrapyard in the White City area of Manchester by a sharp-eyed MG employee from Abingdon. All true sports car enthusiasts owe a debt of gratitude to this unnamed MG employee who, thanks to this devotion to the firm, salvaged this important item of British motoring history.

With the exception of 'Old Number One', all the early 14/28 type Bullnose cars and the later flat-radiator 14/28 and 14/40 types were powered by what became known as the Hotchkiss side valve or L-head engine. This engine evolved from the American Continental Red Seal engine designed around 1913, which Morris was importing prior to the First World War. In 1919 he

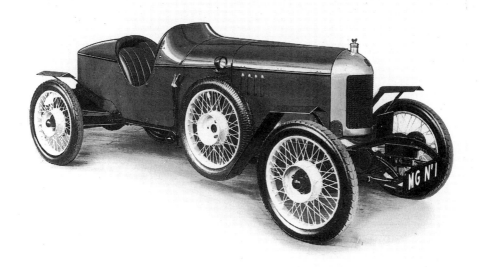

The famous two-seater sports car, 'Old Number One', with a view of the
'works'. This engine was the same type as used in the Glasgow-built
Gilchrist cars in 1920–23.

acquired the manufacturing rights of the engine and reached an agreement with the French armaments manufacturer Hotchkiss et Cie for them to produce it at their factory in Coventry.

Two main changes were made to the design from that of the American-built engines. First, the Ferodo dry plate clutch was discarded and replaced by an oil-immersed cork-lined twin clutch unit. Secondly, since all the Hotchkiss factory machine tools were equipped with French metric tool heads, all the engine and gear box threads were changed to metric whilst retaining British Whitworth spanner size nuts and bolt heads. This anomaly was to cause consternation to virtually all Morris and MG owners for many years, until the practice was abandoned in the late 1930s. The manufacturing agreement lasted until May 1923, when William Morris acquired the Hotchkiss factory and renamed it Morris Engines Coventry.

Apart from all the early Morris and MG cars, plus various commercial vehicles, the engine was produced for marine and industrial use for many years. It most probably continued in some form of production until the late 1950s.

The engines of the Kimber-designed cars differed from the Morris in that Kimber insisted that all the engines fitted to MG cars were stripped, ports and cylinder heads polished, and modified valve guides fitted, together with stronger valve springs. A redesigned exhaust manifold, incorporating a forward exit together with a larger exhaust pipe and extra expansion chamber, was also fitted. All the components were meticulously reassembled and, following a running-in period of 30 hours on coal gas, using a special carburettor, the cylinder head was again removed for decarbonizing, all engine valves being removed and reseated before a Solex carburettor was fitted and the whole assembly checked at a final inspection.

Cecil Kimber, from his earliest days as manager of Morris Garages, seems to have inspired his workforce into paying great attention to detail. This in turn led to a loyalty not just to Kimber but to the product itself. Each worker was regarded as a craftsman and, as if to emphasise the hand-built method of production, it was said of the Abingdon factory shortly before it was closed in 1980 that the only mechanical equipment used was a wheelbarrow!

The 14/28 Bullnose MG, of both two- and four-seater type, had a two-tone body finish of polished aluminium side panels, painted mudguards and body top, usually red or blue. Its modestly tuned engine gave a better than average performance and it proved to be a very good seller, despite its comparatively high price in what were then lean times.

In late 1926, Morris, who had been producing Cowley and Oxford cars with only detail changes, decided to introduce a new wider chassis with improved half-elliptical rear springing but leaving all the other main components almost unchanged. Most noticeable to the general public was

The first production MG—a 14/28 four-seater sports model that was first registered on 31 May, 1924. Note the narrow Bead Edge tyres and the complete absence of any front brakes! (EMGS photo)

The MG assembly line at Edmond Road, Cowley, in late 1927 with several 14/40 models under construction. (EMGS photo)

the disappearance of the round-fronted radiator; Morris never referred to their cars as Bullnose type.

In its place came a flat, upright style, its shape being the nearest any motor manufacturer had yet come to copying that of the very well-respected car from Derby. It was even manufactured from German silver.

Kimber immediately set to work and produced both two- and four-seater body designs to fit the wider chassis, yet still incorporating many of the features that had brought him success in the past, including two-tone body finish, lowered suspension and improved brakes. The early production of this model was still known as the 14/28, but during 1927 it became 14/40.

The fact that a number of other British car manufacturers—Humber, Vauxhall, Bean and Star—all marketed cars under the 14/40 type number, i.e. 14 horse-power RAC rating for road tax purposes and approximately 40 brake horse-power, probably influenced Kimber to change the designation, particularly in view of the MG's performance against the others. In addition to the change of title to 14/40, the radiator badge was changed from the round one bearing the words 'MG Super Sports Morris' with a small octagon MG emblem in the centre, to the now familiar MG Octagon badge.

It was during this period that MG received the unexpected boost of their first recorded race victory, when in October 1927, at the San Martin race track in Buenos Aires, Alberto Sanchez Cires won the one-hour race in his

MG 14/40 at an average speed of 62 mph, and beating, amongst other makes, an Alfa Romeo.

This first win followed a move in September 1927 to a purpose-built plant at Edmond Road, Cowley, this being the first real MG car factory. It must have been a proud moment in Cecil Kimber's career when, in the spring of the following year, the MG Car Company was formed and started to build cars bearing its own guarantee plate and production number.

The new factory continued to produce the 14/40 MG with four-seater saloon and two- and four-seater open tourer bodies, but a move to independence came with the introduction of the MkIV 14/40. This was assembled from individually purchased major components, rather than starting with a rolling chassis delivered from the Morris works.

The Kimber policy of stripping, inspecting and modifying all the main components continued, as did a rolling road and test run before the chassis were driven up to Car Bodies at Coventry, for the Kimber-designed bodies to be fitted.

As if to emphasize this new independence, the MkIV body was embellished with no less than 19 MG Octagon badges plus a further nine octagon shapes if one includes the dashboard and instrument cut-outs, scuttle ventilators, etc. Kimber was now determined to display the brand image of the product from the new factory for all the motoring public to see, proving that he was not only a clever engineer, but an excellent salesman.

Meanwhile the successful William Morris had, by early 1927, purchased the bankrupt firm of Wolseley Motors. A major reason for the failure of Wolseley had been the crippling cost of developing both a four- and six-cylinder overhead camshaft engine. Morris did not use the Wolseley six-cylinder engine design, but later in 1927 did produce a Morris six-cylinder car with an overhead camshaft engine. The car was not well received by the motoring press and, following disappointing sales, was soon discontinued.

Kimber obtained one of these cars but, following an inspection and test, decided to use only some of the components to develop a new six-cylinder MG. When it did eventually appear, in 1928, the car contained virtually nothing of the Morris Light Six design. Even the cylinder block and head had been redeveloped to the extent that new castings had to be made. The chassis was completely new, it being the first that Kimber had ever designed, but he did utilise various axle parts from the 14/40 series cars.

Various forms of both two- and four-seater bodies were developed, and at its debut in the 1928 Motor Show the new MG-style radiator with its vertical centre strip gained motorists' approval. Little did anyone guess that the design would remain in production for more than 27 years, or that generations of small boys all over the world would have ambitions to drive a car bearing the now familiar radiator.

The first six-cylinder cars, known as 'The Quick Six', but later marketed as the 18/80, soon gained an enviable reputation as competitive sports cars in motoring events as diverse as the Monte Carlo Rally, Land's End Trial and various speed and hill-climb events both in the UK and on the Continent.

Alongside the six-cylinder development, Kimber was well aware that William Morris was to introduce a new small car. Morris had seen the sales of his Oxford and Cowley range start to decline in the face of stiff competition from the increasingly successful Austin Seven, developed in various forms since 1922 and selling to an income group that previously had enjoyed travel on only two or three wheels.

Morris used the Wolseley 847 cc-designed overhead camshaft engine in a traditional chassis frame, three-speed gear box and quarter-elliptical springing front and rear. Also, for the first time on a Morris car, an open prop shaft drove the rear axle, fitted with fully floating half shafts. The lightweight ash frame body was covered in leathercloth, whilst the seat cushions were inflatable.

Morris named his new small car the 'Morris Minor'.

The first examples proved to be too lively in performance, so it was decided to market the car with a slightly de-tuned engine by modifying the camshaft timing.

One can imagine the delight on Cecil Kimber's face when he had the opportunity to inspect and drive one of the early Morris Minor cars with the original very lively engine.

He quickly turned his attention to producing a sports car body to fit a slightly modified chassis and used his familiar formula of lowering suspension and steering column to accommodate a lower body line. The body itself consisted of an ultra-lightweight ash frame, covered in plywood and leather cloth. The spare wheel was mounted internally in the pointed tail, and front and rear mudguards were of cycle type. Weather equipment consisted of a two-piece windscreen, with the detachable hood stowed alongside the spare wheel when not in use.

With only two weeks to develop and construct the cars before the 1929 Motor Show, little wonder it was afterwards rumoured that only one of the two cars on display had an engine under its bonnet. It was the first time MG Cars had a stand of their own. Alongside the two new 'MG Midget' cars, priced at £175, stood the also new six-cylinder 'MG 18/80' at £480. The MG stand was the sensation of the show and the 'MG Magic' was beginning to work.

For Cecil Kimber, the 'M' type, as it came to be known, was to prove to be the key to even greater success. Variations of the new small sports car were to go on to win very many races, including the Irish Grand Prix and the Brooklands Double Twelve, so named because the race was run as two

The march of progress—the assembly line at the new MG Works at Abingdon with 18/80 chassis under construction. (EMGS photo)

The new MG Six, later known as the 18/80, being prepared for the 1929 Motor Show at Olympia—the first exhibition at which the newly formed company had its own stand. (EMGS photo)

consecutive 12-hour races. (The local authority would not sanction night-time races because of the noise. Even in daytime, all cars racing had to be fitted with a silencer complying with the local authority regulations; this later became known as the Brooklands Can.)

By mid-February 1931, a supercharged 750 cc version of the 'M' type, driven by Captain George Eyston, was the first car in its class to exceed 100 mph.

Demand for the 18/80 and 'M' type Midget increased as 14/40 sales fell away. The factory at Edmond Road became inadequate and the move to the much larger premises of the former Pavlova leather works at Abingdon took place in late 1929.

It was here at Abingdon that the other models of the 18/80 cars, including the MkII version with its wider chassis and the MkIII racing version with its dry sump lubrication, were developed alongside the many 'M' type variants for both racing and general sports motoring.

Thus we come to the end of the Vintage Years of the production of the early MG cars, with Cecil Kimber and his team of skilled craftsmen well established to make Abingdon the mecca of motoring enthusiasts of several generations from all over the world.

Many of these enthusiasts return in the very cars, all lovingly restored, that were originally built in those Vintage Years.

*   *   *

My own interest in MG cars dates back to the time when, as a youth, I pressed my nose against a car showroom, alas now demolished, to gain a better look at the newly introduced 'J2' type MG. Its twin aero windscreens, large display of instruments and switches filled me with a longing and interest that have remained with me to this day.

Following some motor cycle experience in the immediate postwar period, I acquired my first car, a 1930 'M' type MG. My motor cycling waterproofs came in useful again, since I found it virtually impossible to satisfactorily drive an 'M' type with the hood up, even supposing that, with the thing erected, one could actually get in it. But it was all wonderful fun.

Having been well and truly bitten by the sports car bug, the 'M' type was forsaken for a 1934 'PA' MG. That car gave me (almost) the best motoring years I have ever enjoyed.

Sports car motoring in the 1950s, with uncrowded roads, parking wherever one wished, petrol at six or more gallons for one pound, new tyres costing less than a fiver and seemingly endless blue sky, was a paradise that we now look back on with increasing nostalgia. Because of a chronic lack of spares, many of our repairs consisted of bits from the nearest scrap yards, which were then overflowing with unwanted cars dumped there during the War.

For many years I drove my 'PA' the length and breadth of the land until the need for an additional seat caused me to part with it. Oh, BMA 598! Whatever became of you?

Various other models of MGs amongst other makes of sports cars followed over the years. Then, early in 1987, my eldest son, Neil, another devotee of the marque, told me he had seen, as he put it, the oldest MG imaginable. This I had to see, and being informed it was languishing in an outbuilding in deepest Suffolk, I was soon heading there.

The car, later to be identified as a 1928 MkIV, was the first truly vintage MG I could recall. Although its state of dilapidation was considerable, the idea of rebuilding it appealed to me, provided it could be purchased. My retirement was looming, and I could have spare garage space if I parted with my summer weekend car, a 1500 cc MG Midget.

After some months of inquiries, via my son, regarding the possibility of purchasing the car and badgering him to call more often on the owner, I was invited to pay another visit to Suffolk and open negotiations. My second viewing of the car made me even more aware of the task of rebuilding. Mechanically it looked sound, but the entire body left much to be desired. Even so I was determined to have a shot at it, so a deal was struck. It was not exactly a bargain at the time but, looking back, it was still a satisfactory deal.

After another trip from Lichfield to Suffolk, this time with a low loader trailer hired for a couple of days, I returned home with my newly found MG—or what there was of it—plus various boxes of assorted unidentifiable bits. It is a pity I did not have a camera handy when I parked it on the drive—not to photograph its arrival, but to take a snap of my wife Elma's face when she saw what I had bought.

Having been given the address of the MG Vintage Register by the previous owner, I wasted no time in contacting the then secretary, Phil Jennings. Apart from proving to be a source of much information, including the fact that it was probably the oldest of the only four remaining two-seater MG MkIV cars in existence, he also provided me with two small photos of a similar car in the York Museum, Australia. The National Motor Museum at Beaulieu also provided me with two photographs of the same car at Christie's auction prior to it being shipped out to Australia.

Armed with this enthusiasm, I removed all the body panels, consisting of various thicknesses of aluminium held together with slotted angle iron and an assortment of tin tacks. All this debris, plus the many boxes of assorted bits, resulted in a very overcrowded garage and, since my enthusiasm for gardening was slipping even further down my list of priorities, the garden shed became my spares store.

With the car now in skeleton form, a detailed inspection of its sturdy chassis and axles assured me that all was satisfactory in that department. Since I felt competent to tackle the other mechanical parts, I commenced work on what I guessed would be the most difficult part of the restoration.

The rear section of the duck tail ash frame was altered and the dicky seat narrowed until it looked correct when compared with the photographs. A new floor was made up, which was essential to give me a firm base line to work from. Apart from a few minor repairs to other parts of the ash frame, the forward sections required little attention. The door frames and king posts were original, requiring minimum repair work, and the bodywork forward of the doors presented little difficulty since it is virtually identical to the four-seater version and by this time I had been able to meet owners of four-seaters and make a few notes.

Having obtained an ample supply of suitable aluminium sheet, repanelling commenced, but not before having a word with a skilled sheet metal worker to inquire about the likely snags to be encountered. The problem was remembering all his advice.

I had already decided to retain the scuttle top, so a start was made on the panels on each side of the scuttle, since they have only a single curve. All panel edges were annealed before flanges were turned over a wooden former, extreme care being taken not to mark the outside of the panels as when completed, these were to be polished. The annealing was achieved by heating the panel edges with a blow lamp until a black mark was left on the metal when rubbed with a spent match stick. Who needs a pyrometer?

The rear, top and side panels were fitted next, no great problems being encountered provided each was measured twice and cut to size once. By the time it came to panelling the doors, my fingers were devoid of fingerprints due to handling the hot aluminium. Gloves were soon discarded as useless when trying to handle small panel securing pins.

The door skins, having a compound curve, had to be annealed all over and then flanged and crimped into position while still hot, not an easy task. The encore to this act came when attempting to make the second door look a match for the first one. This was eventually achieved but took a little longer than the first door—two days longer, in fact. This was followed by repanelling the dicky seat lid, where a similar technique was called for.

By now it was winter and I retired to the spare bedroom to carefully unpick such parts of the upholstery as still existed in order to plan its reproduction. It took some considerable time to locate a supplier of suitable leathercloth in the correct shade of dark red and almost as long to pick up the courage to start cutting it into the required shape.

A visit to a local sewing machine shop with a sample of the material provided me with not only the correct needles and thread but plenty of sound advice on how to tackle the job on a domestic sewing machine, which I was

Cyril Mellor's MG 14/40 before restoration began—and a rear view of XV 9508 in all its glory!

assured would be well able to cope with the material. The advice must have been well founded, for during the whole of the task of remaking the upholstery, hood and side screens, only two of the special needles were broken.

As the days grew warmer, I returned to the garage, by this time renamed by my wife as my play pen, checking and overhauling the mechanical parts as required, including clutch, engine, dynomotor and steering box. On its sixtieth birthday I was able to start the engine and reverse the car out of the garage and back again. It had no radiator, exhaust pipe or mudguards, but at least it had moved under its own power for the first time in around 30 years. As I packed up my tools that day, I felt well contented with my efforts thus far.

The owner of another four-seater 14/40 that had been languishing in a field for many years was kind enough to lend me its dashboard and instrument panel to copy. A set of black-faced clocks and gauges came from various sources including autojumbles; not all are strictly correct in every detail, but they will generally pass to all but the most fastidious.

Of the mudguards that came with the car, the rear were correct, but the front were a bit of a dog's dinner. New inner panels had to be fabricated and the rear section reshaped, most of this reshaping and tinbashing being done over the spare wheel rim, tack welding bit by bit until a satisfactory shape was achieved. Once again, the second one took a lot longer than the first to complete in order to obtain a matching pair.

The bonnet top was retained but the hinged side panels were far too tatty and heavily oxidized, so new panels had to be constructed. The side louvres were formed by a local sheet metal shop who also formed for me the 8-foot long side valances, these being the only contracted-out jobs during the restoration. Next the exhaust system and running boards were made up and fitted. Rewiring proved a long task since I wished to incorporate flashing direction indicators within the original side and tail lamps.

The hood posed a bit of a problem for some time, until I was able to locate a photo of one of the other two-seater cars with the hood erected. This particular car has been in Southern Ireland since 1936 and is in very original condition.

It was back to the hacksaw and welding plant again to restore what was left of the hood frame, then up to the spare room once more for a bit more sewing, only this time with a considerably heavier material to sew. Here again the old adage of measure twice and cut once was the order of the day, plus the added precaution of first making a hood pattern from an old bed sheet.

Then came, for me, the least enjoyable part of the restoration, that of painting the bodywork. A long and tedious task, it usually amounted to about five hours rubbing down to every one hour spent actually painting. Still, it

One of only two of the five Mark III 18/100 MGs known to exist—both
owned by Chris Barker. The 2½-litre Mark III was often referred to as the
'Tigress' because of its mascot—mounted here on the front apron—which
was almost identical to the much later Jaguar emblem! (EMGS photo)

all seemed well worthwhile when I surveyed my handiwork.

August 22 1989 was a memorable day (apart, that is, from being my wife
Elma's birthday). It was our very first trip out and we won first prize at our
village show on the hottest day of the year.

To date, XV 5908 has carried my wife and me on many adventures. I say
'adventures', for how else could one describe a drive in a 61-year-old vintage
car from New York to Vermont and back? The 1991 Rally of 'New England
in the Fall' is an everlasting memory.

A couple of rallies around Holland involved driving along the older-style
dykes with their picturesque villages and windmills, all with scarcely the need
ever to open the tool box. The scores of events run by the various MG, Morris
Bullnose and vintage car clubs we have enjoyed these last few years, plus the
scores of new friends we have made worldwide, have proved that 'The Magic
of MG' is still working.

The Lure of Speed

Cecil Kimber

Racing sports cars became an integral part of Cecil Kimber's ambitions once his embryo company was established. He was anxious to test his cars against the other marques and sensed that the results would be crucial to the ultimate success of MG. In the sixty years since Kimber began racing his prototype cars, many thousands of words have been written by others about his achievements. Here, though, is a unique account by the man himself, written in 1944 just a few months before his tragic death. It is part of a lecture he gave to the Specialloids Sports and Social Club in Finchley, London, and the original typescript was thought to have been lost until it was discovered just a few years ago in a relative's attic. It is appearing here in book form for the first time, and offers a unique insight into what inspired Kimber to go into racing as well as an idea of the very modest equipment he had at his disposal. He also offers some highly prophetic views about the racing car of the future . . .

My personal connection with racing really started in a somewhat indirect way during the First World War. I was then working for the Sheffield-Simplex Motor Company and was the owner of a 14 hp T-headed Singer which had raced at Brooklands in the hands of one of the early aviators called Vivian Hewitt—but with what success I do not know. I believe it was capable of lapping Brooklands round about 80 mph and needless to say at the time was the pride and joy of my heart. It was somewhat unique in as much as it was one of the only cars of its day that I ever came across that had a geared-up fourth speed and, as the engines of those days were not given to revving smoothly at high speeds, to click into this high fourth gear made for quite pleasant motoring.

My first active association with racing was round about 1929 or early 1930

Cecil Kimber with his T-headed Singer 14 on the day of his marriage to Renee Kimber in September 1915.

when I was producing the 18-hp 6-cylinder model known as the MG Mark II. I got the Morris Engine Company at Coventry interested in the idea of producing one for racing purposes, and one was duly built and entered for a long distance Brooklands race driven by Callingham of the Shell Mex Company.

The Engine Company were naturally somewhat jealous of their knowledge of the engine and it was therefore left that they should be responsible for the power unit whilst the MG Car Company produced the chassis. Anyway, in those days we did not pretend to have any real knowledge of racing engines, though the engine factory did. The results were disappointing, as the maximum speed from this engine of under 2½ litres was far short of other racing cars of similar capacity, but what we learned from this particular lesson was the necessity of having a fully balanced crankshaft. We did timidly suggest to the Engine Works that this might be necessary, but were told to run away and look after the chassis we were building. The fact remained that, owing to the crankshaft being unbalanced, the throwout loads at high speeds were so great that, when eventually taken down after the race, the crankshaft main journals were actually blue from the heat that had been generated. Needless to say, under these conditions the big-end bearings did not stand up. However, used on the road purely as a sporting car this particular engine gave every satisfaction and had an extremely long life, but it just was not capable of any sustained power output.

Round about 1928/29 someone entirely outside the influence of the Morris organisation at Cowley designed the overhead valve Morris Minor. I immediately saw the possibilities of this little job and designed a light fabric-covered two-seater body for it and, with a few minor modifications to the engine, produced it as the first MG Midget, known as the 'M' type. With good power/weight ratio and an 850 cc engine it had exceptional acceleration performance. It was three of these models that six enthusiasts entered in the Double Twelve Hour Race, and with which they brought home the team prize.

The general preparation work for this race, and the experience we were gaining all the time, increased our knowledge to an immense extent. We had no test shop facilities at all; not even a Water Brake. All our testing was done either on the track or on what we termed a comparator. This consisted of two pairs of pulleys about the diameter of the road wheels, carried on a pair of shafts, one of which was extended and carried a simple air paddle as a brake. The whole was let in flush with the floor, the rear wheels of the car placed on the rollers and there you were.

After this initial success, the petrol and oil interests and the accessory firms began to gather round and make attractive offers which led to a series of racing successes and record breaking that has not been equalled by any other single make of car.

One of the first lessons we had to master was how to make the big-ends stand up. The original engine had connecting rods of parallel section and it was common to find the letter H of the section reproduced on the top half of the big end bearing. Splaying the connecting rod at the bottom end and making it really massive was one step. Abolishing the absurd nick, that is so often milled in the connecting rod to take the big end bolt head, helped; but there was one thing we established and that was the oil temperature in the sump must be kept below 85°C however hard the car might be driven around Brooklands. Very large capacity sumps, well finned, and ample pump capacity provided most of the answer.

Then I designed what became the Standard Midget and Magnette frame for many years, with parallel side members, tubular cross members and an underslung back axle. This at once transformed the car from a road-holding point of view.

The three-speed standard gearbox was then discarded and a 4-speed ENV box with remote control fitted and by degrees it became quite a motorcar.

The first ever race won by an MG was on October 10, 1927. This was in

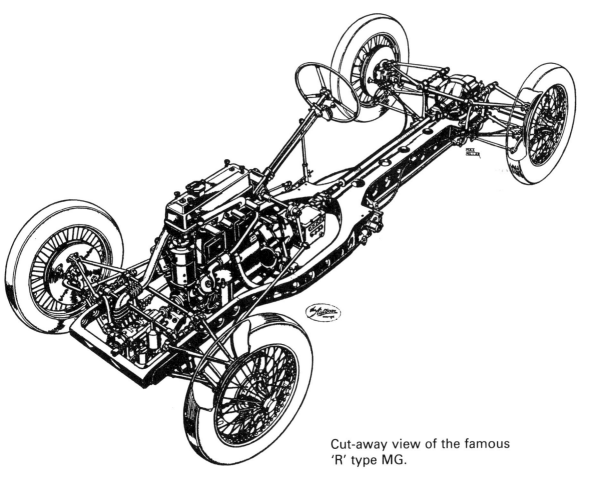

Cut-away view of the famous
'R' type MG.

Buenos Aires, when a certain Alberto Sanchez Cires won a 100 kilometre race on a new concrete track (1½ miles to the circuit) that they had just constructed. He apparently ran away from the field to win at the modest speed of 62 mph.

Over here, the first sporting event ever won by an MG was achieved when I won the gold cufflinks that I still wear by successfully getting through the 1925 London–Land's End without losing any marks, in the first MG that I ever built.

In the following years more similar awards were won, but what really established the MG in the eyes of the sporting fraternity was the way the four 8 hp Midgets entered the 1929 London–Land's End run, sailed up Beggars' Roost in a procession, making faultless climbs and going on to win four gold medals. You must remember that in those days Beggars' Roost failed the majority of entrants, especially in the lower powered range. This was followed by an observed 100 ascents non-stop of this hill.

It was these successes that began to make people think about racing MGs and in time resulted in the record-breaking achievements of George Eyston and Major Gardner which are now part of motor racing history and undoubtedly put the car on the motoring map.

*   *   *

It is possibly interesting to note that the same crankshaft, connecting rods, pistons, valves and cylinder head were used for our racing jobs as were sold to the public and it was only when we came to record-breaking events that any departure from the standard was made. For instance, the car with which Major Gardner broke the World's Record for 1100 cc size engines at 207 mph had a modified but otherwise standard chassis with the engine set at an angle in the frame to bring the propeller shaft diagonally across to one side of the back axle, thus providing a very much lower seating position for the driver. Other than this, the springs, front axle and steering were strictly standard.

As regards the engine, this was a standard K type engine, but had a bronze alloy head with no gasket. The crankshaft was a special short throw one fully counter-balanced, but the connecting rods were perfectly standard, as were the pistons. The inlet valve was standard but the exhaust valve was special in so far as it was sodium cooled. Likewise, the camshaft and camgear were generally standard.

It was rather interesting to note that the horse power developed by the various record-breaking engines always corresponded very closely to the speed in miles per hour achieved and that when George Eyston obtained just over 100 mph the engine was giving just over 100 hp and likewise, when Gardner achieved the 207 mph the maximum bhp of the engine was round about 209, which incidentally I think I am right in saying is the greatest hp per litre that has ever been achieved.

When in 1935 the MG Car Company was merged into the Morris Motor Group and so lost its independence, the four-cylinder 'P' type and the six-cylinder 'K' type counterpart were discontinued.

All jigs and tools were scrapped and we had to get busy trying to make something of an entirely unsuitable push rod job. To my mind this was nothing short of a tragedy and if anyone was to pick up those two engines and develop them from the point we left off, they would have something good to start on.

Concurrently with the 'P' series we produced a pure racing job, super-charged, known as the 'Q' type. This was sold to the public as a regular model ready for the starting line. The 'K3' Magnette was equally purchasable

George Eyston in EX127, with Bert Denly holding the cockpit cover which went over the driver's head.

by all and sundry and this model, in various guises, had a larger number of racing successes to its credit than any other make of car in the world. This may surprise some people, but perhaps it is not to be wondered at when it won major racing events almost without being noticed.

Incidentally, it was the special 'K3' with the diagonal shaft drive and offset differential with which Captain Eyston won a long distance race at Brooklands—I believe it was an Empire Trophy—which was subsequently built into Major Gardner's record breaker.

Following on the 'Q' type came the 'R', which was a distinct break away from MG practice as far as chassis design was concerned. This consisted of a large square sectioned frame shaped like a tuning fork, the engine being placed in the V of the fork. It had torsion bar suspension all round with double wishbones supporting each axle. What, however, was the biggest departure from accepted racing car springing: the fronts had an amplitude of about 4½ inches and the rear 5½ inches. When it is realised that to obtain controllability at speed with conventional springing meant limiting axle

65

The body frame for EX127 being made in MG's experimental department.

movements to about 1½° at the front and 2° at the rear, with very powerful friction shock absorbers controlling the movements, you will realise how revolutionary was this design.

The outstanding feature of this car was the way in which directional stability increased as the speed went up. Anyone who has taken a small car round Brooklands at any speed will know how exciting it became when the 100 mph figure had been reached and what judgement is needed coming off the Byfleet banking. With the 'R' type one had a really comfortable armchair ride with an intense feeling of security. My own personal experience was that very much higher speeds than the car was capable of would have been perfectly safe. Where this design failed was in road racing, as the wheels folded over on a corner. It was precisely this shortcoming of independent suspension on all four wheels that finally brought the Grand Prix Mercedes and Auto Union to the De Dion-type back axle in which the two rear wheels are connected with a beam axle which performs no other function than to keep the rear wheels square with the road and thus stabilise the whole chassis. This tuning fork type of frame would be worth further development, I think, as it lends itself so readily to torsion bar suspension with the torsion bars carried lengthways on the chassis and so not limited in length. Modern air-

66

craft welding techniques could be used to advantage in a frame of this description.

Now a word about superchargers. If the plug manufacturers can, in the future, give us a plug that will stand up to full power conditions and not oil up at low speeds, as I understand they will be able to do, as a result of their War experience, then I think there is a great future for supercharging. The Aspin engine with its screened plug shows another way of overcoming this bugbear.

I, myself, had the most pleasure from a car I owned once. This was a 'K' type Magnette with normal compression ratio and a Rootes-type blower giving about 6 or 8 lbs boost. The normal compression gave good performance at low engine speeds before the effect of the blower came into operation.

This car, a folding head coupé, with four up would do an actual 104 mph on the road. I ran this car for nearly fifty thousand miles and it quite converted me to supercharging for ordinary use. A previous experience with a 1½ litre Alfa in 1929 was equally pleasing.

For racing, however, we used the eccentric vane type of blower using the Power Plus, the Centric and McEvoy Zoller. They absorbed more power than the Rootes type, but the ultimate gain was greater.

Goldie Gardner with his Railton-bodied MG which broke the 200 mph barrier in 1951.

INTERNATIONAL CLASS "F" RECORDS FOR BRITAIN

Lt.-Col. "GOLDIE" GARDNER

with his (MG) regains world speed records★

★ **137·4 m.p.h.**
ONE HOUR

ALSO HOLDS

204·3 m.p.h.
FLYING KILO

203·9 m.p.h.
FLYING MILE

200·6 m.p.h.
FLYING FIVE KILOS

Subject to official confirmation.

Lt.-Col. "Goldie" Gardner also achieved the following records :—

STANDING START
★ 50 KMS . 127·8 M.P.H.
★ 50 MILES 132·0 M.P.H.
★ 100 KMS . 130·6 M.P.H.
★ 100 MILES 136·6 M.P.H.
★ 200 KMS . 135·1 M.P.H.
(Subject to official confirmation)
All these records were attained with a standard T.D. Engine (Supercharged) as fitted to the current production model of the popular M.G. Midget.

Safety MG fast!

Apart from these latest successes Lt.-Col. Gardner's M.G. holds World Records in International Classes "G," "H," "I" and "J" for Flying Kilo, Flying Mile and Flying 5 Kilos. *(Subject to official confirmation)*

THE M.G. CAR COMPANY LIMITED · SALES DIVISION · COWLEY · OXFORD

London Showrooms : University Motors Ltd., Stratton House, 80 Piccadilly, W.1.
Overseas Business: Nuffield Exports Ltd., Oxford and 41 Piccadilly, London, W.1.

Speaking entirely from memory, I believe they absorbed something like 50 bhp at full power. As to the future racing car, I would suggest this should be a 1500 cc machine with eight cylinders in line, or in two banks of four. It was the latter type that the Mercedes concern was developing just before the War and it may interest you to know that in effect it was two 750 cc MG engines on a common crank. This came about through a German racing driver named Kohlrausch who broke records in an MG Midget in Germany— the speed was 147 mph—and whom we presented afterwards with the car as a reward. This car he subsequently sold to Mercedes for 10,000 Marks.

Reverting to this 1500 cc engine, I am of the opinion that the peculiar lozenge-shaped head of the MG, with the valves inclining slightly inwards, had a certain hidden turbulence effect, and although a number of knowledgeable people designed spherical heads with valves at the conventional racing angle of 90°, none of them were as good as the lozenge shape. Personally, I imagine that with modern high compressions obtained by excessive doming to the piston crown, the combustion space becomes in section something like an attenuated crescent moon and the advantages of the spherical shape are, to a certain extent, lost.

Needless to say, I expect the cylinder block will be an aluminium alloy with wet liners and perhaps a bronze head. For the gearbox, something like the ZF, with electrically operated synchromesh gears, would give instantaneous changes. The high octane fuels that will be obtainable after the War will provide one source of increased power, but, to produce a successful racing car, good streamlining and intensely good road-holding qualities are just as, if not more, important.

For this future racing car I have visualised the conventional type of engine, but the new form of supercharged two-stroke with fuel injection into the induction, which is now being developed by Ricardo's, indicates the direction in which greatly increased bhp and MEP will come. Then we must not rule out entirely the petrol turbine, which has made great strides during the War in connection with jet propulsion, though I think such a power unit is still many years away.

The Wonder Years

Barry Foster

The achievements and records that Cecil Kimber's cars set on the world's race tracks fill a book on their own. No surprise, then, that they should be known as 'The Wonder Years', when the various types of Midgets, Magnas and Magnettes took on and defeated all-comers. This aspect of MG history could have no better chronicler than Barry Foster, who has been described as 'the most accomplished club racing driver of the last decade'. Behind the wheel of his 1931 MG 'C' type Midget, VD 30, he is respected throughout the entire vintage sports car movement. Barry also owns the 'C' type in which George Eyston did his practice runs on the Pendine Sands in 1932, before breaking several records in the streamlined Midget, EX-120—and last year he took part in a Diamond Celebration recreation of the event on the same South Wales beach. In the following pages, Barry outlines the history of the early record-breaking MGs and also describes how the Triple M Register of the MG Car Club, of which he is the historian, have been demonstrating all over again what truly remarkable achievements they were . . .

It is easy to allow nostalgia to confound our sensibilities and view the time between 1928 and 1936 as the 'Wonder Years of MG'. Whatever our own personal involvement with MGs produced then or since, a sense of achievement, satisfaction and pride is engendered. However, it is essential not to forget that during these frantic years the M.G. Car Co Ltd was a manufacturing business and had to produce and sell cars to survive. Kimber's genius was to identify a need and provide the means to satisfy it. His maxim of 'make it 5 per cent better and charge 50 per cent more', gave rise to a bewildering array of MG types during those eight years. He also understood that, no matter how good your product, you still had to inform people of what you were selling! MG established a 'house style' of brochures, sales

Barry Foster at the wheel of his 'C' type Midget during his successful record run on the Pendine Sands in September 1992. (Andrew Roberts)

leaflets and advertising that helped to sell the marque. Even allowing for the fact that trading standards were a thing of the future, this material did take some liberties with the truth! Kimber also realised that his cars needed to be seen to be delivering the promises made, and thus took every opportunity to expose MG in the field of combat.

The overhead cam-engined MGs first saw the light in 1928 in the form of a modified OHC Morris Minor. This followed 'Kim's' usual practice of taking one of the Morris models and reworking it with his own body style, coupled with minor chassis alterations and engine tuning. At about the same time, he had fulfilled another of his ambitions, which was to produce MG's own chassis—but still using many Morris and other proprietory components. This was the 18/80 MK2, or 'A' type, and was soon followed by the Mk3 'B' type Tigress. This was a sports racer built for Brooklands along the lines of Bentley, Lagonda and others. A complementary trio of small cars was also produced— in modern terminology these were the first 'modified midgets' to run alongside the Mk3s at Brooklands in the 1930 Double Twelve race. This was 24 hours of racing spread over two days, so that the residents at Weybridge could sleep at night and still allow an event of 24 hours' duration. The Mk3s fell by the wayside, but the 'M' type Midgets won the team prize. It did not take

71

long before replicas of these team cars were available from the factory—at a premium price!

The standard 'M' type Midget was outselling the bigger car by over ten to one. More team replicas were made and even a small coupé was available. Engine development increased the power and it was soon apparent that performance was limited by the Minor-based chassis. Thus by the end of 1930 the scene was set for MG to enter the field of record-breaking.

Records were divided into groups dependent upon engine size, and covered different distances and times. The MG Midget had an engine size of 847 cc, thus falling into class G for engines of 751 cc to 1100 cc and giving itself a large handicap. So when the new record-attempt car was produced, its engine was reduced to below 750 cc to enter class H. At this time class H was the stamping ground of the Austin 7, MG's track rival, and the Ratier and Ridley cars. George Eyston and his friend J. A. Palmes (who sold MGs as a business) were looking for a small car to go record-breaking. After considering other possibilities, they ended up at MG and, with backing from Kimber, the first MG record breaker was born.

Coded EX120 (can one assume they hoped for 120 mph?), the car had a new type of chassis which lowered the centre of gravity, a single-seat streamlined body and a much modified engine using 'M' type and non-MG parts. This was the work of Reg Jackson and Sid Enever of MG and also E. A. D. Eldridge, who was not an MG employee. The race was on to be the first 'baby' car to reach 100 mph. Austin were close and EX120 went a little faster, but not enough. Eyston had an interest in a company making the Powerplus Supercharger, so one of these was fitted and the car taken to the Montlhery Circuit, just outside Paris. Brooklands was closed for repairs at the time and also required silencers to be fitted, which could reduce power outputs. The cold weather caused some problems—it was January—and a radiator cowl was made out of an oil drum, or so we are told! But Eyston went round at over 7000 rpm and 101 mph for ten miles, to be the first 750 car to do the 'ton'.

Kimber exploited this achievement with adverts that simply stated it was an MG Midget—leaving the reader to assume it was very similar to the model on sale to the public. Behind the scenes, MG's next production model had many of the features of EX120, but with a full two-seater body and the MG radiator. This was EX125, better known as the 'C' type, or Montlhery Midget. These won the 1931 Brooklands Double Twelve race, taking the first five places, and later in the year won the Ulster TT race and Irish Grand Prix. Many other successes were achieved by the 'C' types, including some record breaking by R. Horton in a 'C' type with a single seat body of his own design. Eyston also used EX120 for racing, and it was even road registered! However, he had plans for 100 miles *in* the hour. This was achieved in

MG advertising! The 'M' type Midget which accompanied EX120 when it took the 100 mph record.

September at Montlhery, but with the loss of EX120 in the process. Eyston had covered the distance and taken the record, but missed the flag and went on for another lap, during which EX120 caught fire. Eyston baled out and the car came to rest on the far side of the track—but was burnt out. Eyston suffered some burns, though he was not put off record breaking. He did, however, have to delay his first attempt in EX127 whilst recovering from his injuries.

The lessons from EX120 and the 'C' types produced the Magic Midget, coded EX127. This really was the car in which to go for two miles a minute. Six days after the EX120 fire, Eldridge took the first record in EX127 at 110 mph. Three months later, Eyston put it up to 114 mph. However, 120 mph still eluded them. The scene then moved to South Wales, to Pendine, to seven miles of sand, the site of world speed record attempts by Parry Thomas and Malcolm Campbell.

There were advantages and disadvantages to running at Pendine. The crew had to wait for three weeks until the sand was hard, smooth and dry. Salt and sand do not mix with highly-tuned engines and there was always the problem of the tide turning at the critical moment. It did mean, though, that records taken would also be British National as well as international records. EX127 ran at 120 mph but the timing apparatus failed! A rerun with the tide on its way in gave a two-way speed of 118.39 mph. Winter conditions forced any other attempts to be abandoned. It was not until December that EX127 returned to Montlhery and Eyston took the 120 mph record. Later, EX127 with 'C' types number CO254 and CO268, and J3 J3756, took every single class H record—the first, and so far only, time all records in a class have been held by the same marque. One event that could have given MG a unique record was Bert Denly's drive in EX127 at Montlhery. He took the 100 kph record at 132 mph; if he had continued to the hour at this pace it would have been the fasted *any* car had travelled for one hour. Even John Thornley, in his book *Maintaining the Breed*, gives no clues as to what happened.

Meanwhile, back at the factory, MG had produced two other types of OHC car, the Magna and the Magnette. Both became involved with record breaking. The first of the Magnas was a six-cylinder version of the Midget, offered with a variety of body styles, but was not conducive to tuning like its four-cylinder counterpart. I know of no record attempts by an 'F' type. However, the L Magna had the same basic engine as the K series cars and was thus a good starting point. An L2 took class G honours for 12 hours and also 2000 miles at over 80 mph at Montlhery. However, class G became the territory of two K3s—K3007 and EX135.

K3007 was the property of Ron Horton, who had been taking records with his special-bodied 'C' type. He repeated the concept with the Magnette and in 1934 at Brooklands took his first records—some 16 mph slower than EX127 in the smaller class. Horton managed to get more power from his engine than any of the other K series engines. The year 1935 was one of change. MG withdrew from racing following an edict from Morris and were coerced into returning to the use of more Morris parts in their products. The OHC engines gave way to push rod units of the SA, VA and WA types and saw the birth of the new Midget, the 'TA'—which started a whole new era in MG history that took the company on to sales figures unheard of in the OHC days.

Record breaking still continued, however, for racing went on and the OHC cars were modified and 'improved' by private owners, with factory help through the racing department. The old names disappeared, new characters replacing them and making MG even faster. EX127 was sold by Eyston, who

EX127, with the body built specially for Bobby Kohlraush, takes off on one of Brookland's famous bumps!

was now concentrating on other forms of record breaking. The Magic Midget was purchased by the German driver Bobby Kohlrausch and, after some use with the earlier engine, was returned to the works and rebodied with a later 'Q' series engine fitted. This engine is reputed to have produced 150 bhp— all from 750 cc! Using the German autobahn, Kohlrausch upped the mile record to over 140 mph in 1936.

At this time appeared the most successful of the MG record breakers, K3023—better known as EX135. This was a special K3 built for Eyston along the lines of the Magic Midget. It was known as the Magic Magnette— or, less politely, as the Humbug because of its brown and cream stripes! It was quicker than the Magic Midget for many distances, but in Eyston's hands did not better the faster of the Midget's times. Meanwhile, enter Major A. T. G. 'Goldie' Gardner who purchased K3007 from Ron Horton. This was used for racing and some record breaking, equalling the speed of EX135 at Brooklands, and thus setting British National records almost as fast as the international ones. Gardner acquired EX135 from Eyston and had in the

75

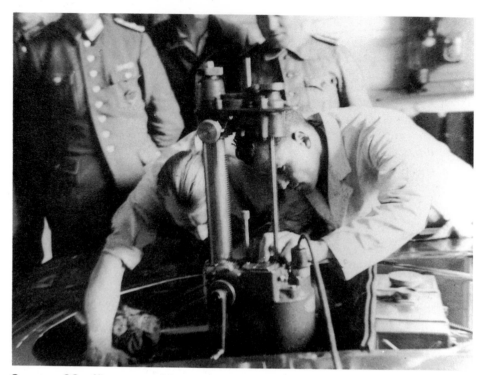

German SS officers watching while MG mechanics Reg Jackson and Syd Enever rebore EX135's engine between record runs!

meantime used K3007 to become the fastest-ever MG, taking over that slot from EX127. At this stage, EX135 became known as the Gardner-MG, as it had received a new body designed by Reid Railton and an engine from K3007. In this new guise, the car returned to the Frankfurt autobahn and raised the record from 148 to 186 mph! Six months later she was back in Germany at Dessau autobahn when Gardner and MG pushed the class G records to over 203 mph. Two days later, EX135 took the class F records at over 204 mph for cars up to 1500 cc. Reg Jackson and Sid Enever had taken a portable cylinder-boring bar with them, plus the necessary pistons. After the 1100 cc records had been taken, they removed the cylinder head and rebored the engine to take it into the next class up in size, by a couple of cubic centimetres!

What is even more remarkable is that legend has it that each thought that the other knew how to work the boring bar—so under the gaze of SS officers and other onlookers they learned how to do it on the record-breaking engine! The speeds achieved next day show that they were fast learners.

And so the War came and the end of record breaking with the OHC MGs

. . . until 1946, when Gardner took his MG to Belgium and on the highway at Jabekke raised the 750 cc record to 159 mph using the six-cylinder engine with a short stroke crankshaft. Two years later he returned with a Jaguar-engined EX135 and took some class E records. He had not forsaken the OHC MG engine, however, he simply took out two pistons and ran it in the 500 cc class at 154 mph. The following year two more pistons were removed, to reduce the engine to 350 cc, and Gardner still cracked the records at 120 mph—a record that the Magic Midget had strained to reach with an engine twice the capacity. But even EX135 was not finished.

The Gardner-MG crossed the Atlantic in the hold of the *Queen Mary* and with a blown XPAG MG 'TD' engine set more records. Another OHC engine was fitted and though this was a modern Wolseley unit, further records were set. Finally, another XPAG 'T' series engine was installed for her last record-breaking session in 1952 at Utah. Here, once more, EX135 again exceeded 200 mph, some 13 years after her first double-ton records.

*   *   *

Just 34 years after EX135 had turned a wheel in anger, OHC MGs were back record-breaking. In 1986, the Triple-M Register of the MG Car Club,

Goldie Gardner tinkers with the six-cylinder engine of EX135 during one of the recording-breaking runs.

Ron Horton in his single-seater 'C' type at Brooklands. No crash helmets were worn in those days!

which looks after the interests of owners of OHC Midgets, Magnas and Magnettes, went into action on the Motor Industry Research Association's banked track. Four cars were prepared for action to attempt to take British National Speed records. The smallest was a J2 with the engine reduced to 500 cc, owned by Mike Hawke—who also owns the ex-Horton K3007 used in pre-war record attempts. Class H for 750 cc cars used my 'C' type, while class G for 1100 cc cars had Peter Green's K3011 and Len Bull's J2. The plan was for three cars to run for distances and records up to 24 hours, with the other J2 going for some shorter distance times. It was unfortunately hampered by a split fuel tank which prevented it obtaining its targets. The K3 set five records, the 'C' type recaptured two records from Austin, and the 500 cc J2 set eight records. All three cars completed the 24 hours— though the 'C' type lost over an hour with an engine fault.

Encouraged by the 1986 results, the Triple-M Register were at it again in 1989—this time at the Millbrook banked track. The K3 and the 500 cc J2 again took part, along with Ken Rees 'C' type and a PB special, bored out to nearly 1100 cc, belonging to Sean Smedly. The J2, 'C' type and K3 were attempting to improve on, and set, additional records to the previous attempt. The PB would be going for the shorter distances. At 7.30 pm the three 24-hour cars started their runs. Pit stops and driver changes occurred at the designated times. The first hint of trouble was a misfire on the 'C' type. Various pit stops to try and cure this eventually put the car too far behind its lap schedule, so in the early hours of the morning it retired. At dawn the PB started its

attempt and after some very quick laps seemed to be on target when problems
in the clutch department resulted in an unscheduled stop. Another Register
member drove some distance home, collected the clutch from his own car
and had that installed. The PB made another start, only to shed the tread
from a front tyre when travelling at over 100 mph! Fortunately there was no
other damage. It was decided not to continue as the safety of the other tyres
could not be guaranteed. The J2 refused to start after a pit stop and required
the inlet manifold to be repaired before returning to the track. The K3 ran
faultlessly for 18 hours until it, too, developed a misfire, diagnosed as supercharger seal failure. The car was stopped and, as it was so far ahead of
schedule, the innards of the blower were removed and the car dispatched
with a very long and peculiar inlet manifold with the carburettor at the end.
It still lapped at over 60 mph, but was now slower than the 500 cc J2. At
the end of 24 hours the two cars claimed two international and 30 British
National records.

In 1992, MG record breakers returned once again to Pendine sands, 60

Barry Foster in his 'C' type wresting back 750-cc records from Austin in
1986.

years after the Magic Midget. John Bannell had put a single-seat body onto his J2 MG and borrowed the 500 cc engine from Mike Hawke's record-breaking J2 for an attempt on the mile and kilometre flying start records for 500 cc cars. My 'C' type, used in the 1986 session, would be attempting the ½-kilometre record. The 'C' type was actually racing in France the weekend prior to the attempt when it broke a crankshaft—still a problem with the two-bearing engine. The car was, though, ready for the attempt in good time—the day before!

The atmosphere at Pendine was MG Magic! The Beach Hotel, used in all previous attempts, was HQ for the weekend. Lots of other MGs took part in a road run and watched the proceedings. Several of the 'Taffiosi' had been there as young lads in 1932. The only real difference was that this time it had to be on Sunday; no waiting for three weeks for perfect conditions—just get in and go! The very modern timing equipment was housed in a replica of the 'M' type high speed van and the practice car was the ex-Hugh Hamilton 'C' type. Eyston had used the works hack—another 'C' type, registered RX8306—to look at the beach prior to EX127's record runs. Also in 1932, to stop the high speed van from sinking into the sand, it was put into first gear, the steering wheel put onto full lock and tied into place with string. Then it was started up and the driver got out and left it going round in circles!

My first runs in the 'C' type were made at about 80 mph, when it was found that, because of the wet sand, considerable drag was experienced. Extra boost was used from the supercharger, causing the plugs to cook. Several hours were spent retuning with harder plugs and a richer mixture until engine guru Bob Jones was happy with the result. At about 4pm—when the tide had turned—the 'C' type made its two runs and set up a mean time of over 93 mph for the record.

The sensation was rather like driving on several inches of soft snow—uphill!—with the ground underneath covered in ice. The ripples in the sand gave the whole thing a high frequency vibration, which left me feeling that any steering I did happened some seconds after I had turned the wheel and that the back end was really doing the steering by wandering from side to side! I have even greater respect now for Eyston, Campbell and Thomas, who all went very much faster than I did.

The 500 cc J2 had some ignition problems but, as it had not been intended to run the car until the following year, its rapid completion and success was a credit to the owner and the driver. The standing start 1-mile record was set at 39.67 mph.

And there the story ends for the moment. Do not expect these OHC MGs to rust away in some dusty garage. Plans are being discussed even now for the next attempt on records—*Laudes Auget Priores*!

Barry Foster warming up his MG Midget prior to his record attempt on Pendine Sands in September 1992. (Andrew Roberts)

Barry Foster at speed during his record-breaking
run on Pendine Sands.

The MG Fast Ladies

Anna Guerrier

Although motor racing has long been considered a very male sport, it is a fact that a number of women have made substantial contributions to it over the years—breaking records and achieving outstanding honours of their own. Nowhere, perhaps, has this been more evident than in the legend of the MG. Indeed, during his lifetime Cecil Kimber did much to encourage women racing drivers, which may go part of the way to explaining the appeal the car continues to exert for the fair sex! Anna Guerrier is in a unique position to recount this often overlooked area of motor sport, being a racing driver as well as a writer and radio and TV broadcaster. She brings considerable experience to her writing as a historic motor sport competitor—having regularly taken part in the MGCC BCV8 championship and won the Jean Denton Trophy on three occasions for the highest female driver. The MG Fast Ladies have a worthy champion in Anna Guerrier!

The needle on the rev counter swayed at 3,000 revs. My hands, clammy, twitched on the steering wheel. My eyes were glued to the lights. On came the red, the wait seemed interminable, then the green. I dropped the clutch, moving forward with a great leap, only to brake and swerve to avoid the car ahead stuck on the grid. Cars were pouring past me. I accelerated again to join the mass of high-speed metal scrabbling round Copse. This accomplished, the pack began to spread slightly as I pulled to the right of the track and kept my foot flat to the floor, drifting through Maggotts. So began my very first race at Silverstone.

This country has a great female motor sport heritage which is unknown to the majority of the sport's devotees. History books document the success of George Eyston, Reg Parnell, the Bentley Boys, Sir Malcolm Campbell—the list seems endless. Women have also been driving successfully since the

Anna Guerrier behind the wheel of her MGB at Snetterton.

'twenties and 'thirties and, though in proportion to the numbers of male competitors they appear to be a meagre few, there are still far too many to mention here.

Motoring in this country began to take off with the building of the Brooklands Motor Course in 1907 by wealthy landowner Hugh Locke King on his estate in Weybridge, Surrey.

He identified the need for a track where cars could be tested, developed and raced if the British motor industry was to keep up with European counterparts. The gates to this legendary 2.75-mile oval race circuit, 100 feet wide with its dramatic banked sections, opened on 17 June. Nineteen days later, on 6 July, the inaugural race took place, with the first ladies' race following only a year later in which Ethel Locke King took the honours.

William Boddy, in his book *The History of The Brooklands Motor Course*, chronicles the ladies' achievements and it is then that you realise just what a prolific and talented bunch they were. It was a regular occurrence for ladies to be seen competing alongside the men or in special ladies' races. It would be foolish, however, to think that women were welcomed to the race track with open arms. Public, officials and male competitors alike doubted the ability of women such as Elsie (Bill) Wisdom to control her huge Leyland. She proved, however, that not only was she able to cope but went on to take the Brooklands Outer Lap Record and in 1932, with Joan Richmond, won outright the 'Double Twelve' (a race of twelve hours per day run on two consecutive days). A continuous 24-hour race had been banned by the local

84

council before the first Double Twelve due to a record attempt by Selwyn Edge of Napier. He drove his Napier for 24 hours non-stop, achieving 1,581 miles, 1,310 yards, at an average of 65.905 mph and establishing a total of 26 records, the World 24-hour record standing for 18 years. The council banned any further 24-hour race after complaints from local residents about the noise.

Other women whose names featured in racing circles were Kay Petre, Gwenda Hawkes, Doreen Evans, Eileen Ellison, Margaret Allen, Fay Taylour, Mrs Tolhurst, Dorothy Stanley Turner, the Honourable Mrs Victor Bruce and Miss Schwedler, to name but a few. Reading reports of the time, these ladies drove with success and panache which gained them recognition and respect from their male compatriots. They also managed to beg, steal and borrow cars in which to compete. For example Kay Petre borrowed Oliver Bertram's 10-litre Delage for her speed duel with Gwenda Hawkes. Fay Taylour borrowed an Alfa-Romeo and Margaret Allen won the First

Cecil Kimber (in trilby) with one of his leading lady drivers, Mrs 'Bill' Wisdom, and her K3 Magnette at Brooklands.

Two more of the early female entrants into the predominantly male world of motor racing were Kay Petre and Doreen Evans.

August Short Handicap in a single-seater Frazer Nash. But MGs were favoured by both men and women and were frequently in evidence on the track and amongst the front runners.

Cecil Kimber, originator of the MG marque, was keen to see his cars performing well on the track and encouraged women to enter races. The Brooklands relay race of 1934 was a prime example.

Three MG Magnettes were prepared by the factory but entered by Miss Schwedler, the rest of the team consisting of Margaret Allen and Doreen Evans. Their goal was, of course, the Ladies' Trophy and the hinted-at possibility of a works drive at Le Mans.

Kay Petre led the rival Singer team. Both camps worked hard throughout the preparation and practice session, scotching theories that female racing drivers meant lipstick and powder puffs and positively dumbfounding those gentlemen who couldn't believe that these women knew a spanner from a wheel brace.

The race started with cars of all different shapes and sizes jostling for position, but before long the MGs were lapping well, with Margaret Allen putting in speeds of over 100 mph, a pace the Singers just could not match.

As the race progressed, Fate played her hand and a gust of wind sent the MG team notes disappearing up the pit lane, where they were retrieved by Kay Petre. As she gathered up the various sheets of regulations she noticed some small print which stated that any crew finishing within the top three would not be eligible for any further prizes. Clearly this was to have an effect upon her strategy for the Singer team. Now there was a choice; unable to come in the top three *and* win the ladies' prize, they would have to drive to win one or the other.

As the race progressed, solid sheets of rain were making driving conditions more than hazardous, but this helped the MGs to creep further up the order as other competitors spun off or broke down. 'Miss Evans', writes William Boddy, 'drove amazingly well through the hail of raindrops' and the MG girls were still going well when they crossed the finishing line in a very impressive third position overall. They had managed an average of 87.86 mph, which is on a par with a modified road-going MGB in one of today's club championship races.

The Singer team, made up of Kay Petre, Eileen Ellison and Mrs Tolhurst, came in fifth, which was enough for them to take the top ladies' accolade, the Ladies' Prize, and to cause some disappointment to the MG team.

The prestigious Le Mans 24-Hour Race has always conjured up an atmosphere of excitement, drawing spectators and competitors from near and far. In 1935, a works team of MGs was entered under cover of George Eyston's name, but all the cars were driven by women. This was a definite first. The team was made up of Margaret Allen, Colleen Eaton, Joan Richmond, Barbara Skinner, Doreen Evans and Mrs Simpson.

There were two categories in the race. The first, for the greatest distance covered, inevitably went to one of the larger-engined vehicles, while the second, the 'Rudge Whitworth' trophy, was run on a handicap system giving the smaller-engined cars a competitive opportunity. To qualify for the 'Rudge', the competitors had to have finished in the previous year's race. The all-girl MG team was not eligible on this occasion, but they were determined to make a 'Rudge' victory possible for the following year.

After the initial mayhem of the start, the MGs settled into a regular pattern. As darkness fell, the rain began to fall in torrents, but on and on through the night the three cars kept exactly to their schedule. As streaks of light in the early dawn appeared, the spectators, who had not considered the possibility of a feminine team lasting, were delighted to applaud in full measure their remarkable endurance.

All three cars finished, having completed 1,567 miles in 24 hours—which was the more remarkable for MG since 30 cars had retired during the race.

Sadly, strikes in France were to mean the cancellation of the race in 1936, so it was not until the following year that the challenge for the 'Rudge' trophy could take place.

Rain seems synonymous with motor racing and certainly the 1937 Le Mans 24-Hours had its fair share. The start took place in the most violent of storms. Lightning maliciously seemed to follow drivers about and heavy rain made it almost impossible to see at speed. The battle for the 'Rudge' was intense and difficult to assess, as a slight increase in speed by one team necessitated an increased average for all, with copious amounts of mathematics necessary to ascertain the correct speed.

Dorothy Stanley-Turner, who had learned to drive at an early age, piloted a little green PB with Enid Ruddell acting as her partner. Dorothy, a family friend of the Kimbers, was a great MG enthusiast. She had first started racing with her own J2, originally built for Le Mans, followed by a supercharged 'Q' type in which she made many appearances at Brooklands and Donnington. But she excelled at road races and competing at Le Mans was the fulfilment of her greatest ambition.

Despite the rain, she and Enid drove steadily and, as dawn approached, it became evident that they were lying in second place for the 'Rudge' behind the Aston Martin of Morris-Goodall and Robert Hichens.

Racing would not be the same without some drama and Dorothy and Enid had theirs when the French *plombeur* who was filling the PB with petrol pulled the lid off in his hand. Strictly speaking, they should not have been allowed to continue, but Dorothy found an orange, rammed it into the mouth of the petrol tank and secured it. After some fast talking, the officials allowed her out onto the track once again, but valuable time had been lost.

Driving to the best of her ability, she scorched past her competitors, finally managing to regain the lost places and eventually catching up with the Aston Martin again.

Le Mans is of importance not only as a race in its own right, but is seen as a showcase for motor manufacturers throughout the world. In '37, Aston Martin's future lay in the balance and it was imperative, if the company were to survive, that their car should finish and make a good account of itself.

No sooner had Dorothy clocked in behind the Aston than nobody saw it circulate past the pits. As the seconds ticked by, the team began to worry and in desperation the team manager approached the MG crew to see if Dorothy would consider stopping to give the missing driver a message.

What a dilemma! With the Aston out of the race, Dorothy and Enid would win the trophy, but Dorothy did not hesitate in making her decision. She found the Aston parked just off the track, stopped, and told the driver to keep going if in any way possible. She then continued on her way.

At the end of the day the MG PB came in second, behind the Aston

Dorothy had so chivalrously stopped for, but it was an outstanding position for the two young women in their first-ever continental road race.

Dorothy had been taught to drive as though she were a boy. As a result she was more mechanically aware than any of her rivals, which was to stand her in good stead as she continued with her inspired racing career. Her Brooklands races were always exciting, as she took all the corners in skids and slides so near the point of no return that after her races, officials and spectators were sent scurrying to the bar to be revived!

Sadly, the Second World War brought racing at Brooklands to a close and with it Dorothy's escapades. Brooklands was turned over to the war effort and the aviation side of activities, which had always been a healthy part of life there. So, Brooklands became the home of aviation much as it had been the home and birthplace of British motorsport.

Immediately after the War, the opportunity to indulge in circuit racing was not so easy and more people turned to rallying. Here was a type of competition which offered a variety of different forms within a single event; flat-out special stages, timed road sections, taxing navigational skills and a demonstration of ability to control a car in all different weathers, road conditions and climates.

Nancy Mitchell and her husband started rallying in 1948 on the Alpine Rally in their HRG as part of the team which won first prize. That was just the start. Succeeding seasons saw Nancy, housewife and mother of three, making a name for herself in hill climbs, sprints and rallies all over the country.

She joined BMC in 1956 for her most prolific and successful year. As part of the competition team, she had a hectic year including events such as the Monte Carlo, where she finished third in the Ladies category, Sestriere (2nd) and Lyon (2nd). The Tulip saw her in an MG Magnette, a particularly memorable event as she was chucked out. 'The organiser tapped the bonnet and said in a broad Dutch accent, "This is not metal". Well I knew what he meant, and I threw my hands in the air.' 'The whole bloody thing is aluminium, mate!' was her response and resulted in disqualification and considerable discussion about the 'spirit of the rally'.

She continued the year with the Midnight Sun, the Alpine (first in Ladies Class), the Liège (second in Ladies Class), Viking, the RAC (again in a Magnette) and Lisbon rallies, in all of which she took second place. In the Alpine she was the first team car home in her MGA, 15th overall and winner of a coveted *Coupe des Alpes* for not incurring any road penalties. She competed in the notorious Mille Miglia that year, during which five drivers and nine spectators lost their lives, but Nancy came third in her class and won the Ladies' Prize. The just reward in '56 was being crowned European Ladies Rally Champion, driving an MG on all her points-scoring events.

89

Pat Moss, like her brother Stirling, enjoyed great success on the race track, and is seen here with Ann Wisdom. (Autosport Picture Library)

Another hectic year followed with equally meritorious achievements. Piloting an MG Magnette and an MGA 1500, she proved her worth within the team by winning the European Ladies Rally Championship for the second year on the trot. Nancy's 'never give up' approach, plus her stamina and bravery, which she regards as inherent qualities in all females, meant that the final package made her a force to be reckoned with by any competitor— male or female. One of the team mates of her time with BMC was the highly accomplished Pat Moss. Though renowned for her Austin Healey drives during the 1960s she had started her serious rally career in an MG TF.

Pat originally started to get to grips with rallying in a Morris Minor, but this was getting rather tired, so after some encouragement from her father she bought a Triumph TR2. She wanted to enter the car in the '55 RAC and so approached Triumph to see if they would give her some financial help with her expenses. They were not very helpful, offering instead a car (which she already had) and no expenses. As a result she ventured to ask John Thornley at Abingdon for some assistance. The outcome was the loan of a TF *and* expenses.

The car arrived, minus the basic rallying extras, but that was soon fixed and with Pat Faichney in the navigator's seat they set off as part of the works team for the RAC. Pat was concerned that their rally number 193 added up to 13 and, as if to confirm her superstitions, the car skidded off the road, breaking the headlamps, on the way to the start. However, they made it to Birmingham in time to be flagged away on one of the most gruelling events on the calendar.

Navigational problems hindered their progress, but Pat made up for things on the timed sections. Sheila Van Damm, an experienced rally driver, was most disgruntled to find that Pat was consistently the fastest lady on the special stages, which consequently made her even more determined to find some extra power. She finally beat Pat into third position in the Ladies, which, considering it was the MG duo's first major event, was not a bad result.

The MGA made its rallying debut with Pat at the wheel in the 1956 RAC. Despite numerous excitements during the event, Pat was leading the ladies category by 70 points when approaching the penultimate stage—but a wrong slot on Blackpool promenade meant incurring 100 penalty points, which took her from first to third.

Many of the long events meant that the drivers became both physically and mentally tired. Pat is reported to have said that on her first Liège Rally in '57 she had started to see 'black cats' and 'burning cars in the road'. Caffeine tablets or some form of wakey-wakey pills were quite the norm to keep drivers going through the long and sleepless sections of driving. With night sections increasingly incorporated into Classic and Retro events today, doubtless this will again be the case!

By 1958 Ann Wisdom had become Pat's permanent co-driver. Having come from well-known racing parents, Elsie (Bill) and Tommy Wisdom, motor sport was very much in the blood. This was a partnership destined to continue with great success over the next four years.

By the time Pat had completed several events in '58 it was evident that she was high in the running for the Ladies European title, so it was decided that she and Ann should go to Stockholm for the Midnight Sun Rally with one mechanic and an MGA 1500. The car was not ideal for the event, especially as there were so many Porsches in the same class. After a calamity which saw the car off the road, they finished out of the championship points. The event was not a complete loss, however, as it was on this rally that the girls first met Pat's future husband, Eric Carlsson, when he offered them an orange while waiting at a time control. Despite the lack of points on the Midnight Sun, 1958 was indeed to be Pat and Ann's year as they took not only the European Ladies Touring Car Championship but also came sixth overall in the main European Touring Car Championship.

The Alpine of that year was the first Healey 100/6 drive for Pat. Having had blocks of wood added to the pedals so that she could reach them, she went on to be the highest placed British driver, coming fourth overall and winning the Ladies class. From that time on, Pat divided her drives between

91

the big Healey 3000, an Austin A40 Farina and a variety of other BMC machinery. She achieved considerable success in all of them. During the period from 1959 to 1962, she and Ann won the Liège, the Tulip and the German rallies outright, took second place overall on the German, Alpine, RAC and Polish and third on the Alpine, Geneva and the RAC. Add to that the class and ladies wins and you have a trophy cupboard positively bursting at the seams.

In March 1962 Ann, affectionately known as Wiz, married Peter Riley, thereby uniting two well-known motoring families. By June she had told Pat that she was expecting a baby and therefore had to give up rallying, so ending a happy and successful rallying partnership. That year was also the last competitive year that Pat spent with BMC. Ford made her an offer she could not refuse and so in '63 she made the change. Stuart Turner, who was then running the team, gave her seat to Pauline Mayman, the co-driver who had taken over from Wiz, and signed Val Domleo as her navigator—another partnership tuned for success.

Jean (now Baroness) Denton has an MG association of great renown. She learned to drive at the age of 26 in Trafalgar Square and within one year of having a driving licence was racing with such success that she won the British Women Racing Drivers' championship.

She started off with one of the first Mini Coopers, progressing to a Lawrence-tuned Morgan and then to a Formula 3 Cooper previously owned by Jackie Stewart. Unfortunately, though, the great man had taken the engine home with him. 'The car went round corners fantastically,' she said, 'but everybody overtook me on the straights.' Denton soon realised that motor sport is about money, not cars. With incipient world champions spending £1,000 to get an extra half bhp, she felt the decision to change to a production car on the grounds of cost was inevitable. So began the Denton/MGB liaison.

Having found a suitable 'B' roadster, it was prepared for racing by her husband and Tom Boyce, with advice from the experienced racer Bill Nicholson. The car was first used for a complete season of club meetings in production sports car races before winning the British Women Racing Drivers' Championship in 1968.

From there, Europe beckoned and the path to the Nurburgring, Mugello, Barcelona and Lisbon. She competed in Group III and, though never likely to beat a Porsche on speed, she was always there when the chequered flag was waved, whilst the 911s were not always so reliable.

The high point in this relationship between car and driver came on the London to Sydney Marathon. 'We had learnt a lot about the car. There was no doubt that I had one of the best MGs around, so when the *Daily Express* announced it was organising the first ever marathon to Sydney it seemed to me the one and only way I would stand a chance of reaching Down Under,'

The indefatigable Jean Denton *(above)* on the London to Sydney Race in her faithful MGB Roadster and *(below)* Circuit Racing.

reasoned the irrepressible Baroness.

The 'B' was stripped and rebuilt. Seams were double welded; fuel and brake lines were brought inside; powerful spotlamps and headlamps were added; a double fuel tank took up the entire boot space, while three jerry cans were attached to the back for the stretches in Australia where there was little likelihood of a petrol halt for a good many miles.

The passenger seat was replaced by a bed, which could be propped up sufficiently to navigate from when not sleeping, and under all of this the spares were packed to keep the weight as low down as possible. Strengthening the suspension was no problem and meant that the car was easily able to cope with the rough terrain ahead.

All of this work cost money and Baroness Denton set out to find sponsorship in order to finance the whole undertaking. The response was impressive. The car was entered by *Nova* magazine, the editor of which was heard to say, 'If those two get bloody lost in Dover High Street . . . '

The car was painted to look like a high speed cigarette packet in metallic gold and black and the team (Denton and Tom Boyce) were dressed by a designer from Worth. Other companies pledged considerable sponsorship and there were some small cheques from well-wishers, all of which made the trip possible.

Crystal Palace was the scene of the start and an atmospheric occasion it was. Once across the Channel, they travelled in horrendous fog across France, through the Mont Blanc tunnel into Italy and on to Yugoslavia, the roads becoming rougher as the route took them down through Turkey and Iran. The added weight of their equipment was to cause them additional problems in the snow and ice of the mountain passes, but soon even these difficulties were left behind.

Teheran special control offered the opportunity for Jean to sleep whilst Tom had help with the servicing. From there it was on to Afghanistan, by which time they had lost overdrive, which wrecked their petrol calculations and ruined the top speed. This was a particular disappointment as the roads turned out to be some of the best encountered.

The MGB disappeared down the old silk route to the Khyber Pass in Pakistan and India, where the enthusiasm of the crowds was most notable. Reaching Bombay, the boat section to Perth began and this turned out to be, for Jean, the horror section of the rally.

'The famous Bombay tummy got me and I lay prone for most of the trip. The sight of land at Fremantle was welcome, but was by no means as welcoming as the MG Car Club members who lined the quayside.' From then on, guardian angels cared for them across Australia.

However, it was during this section that problems started to hit them thick and fast. First a fuel pump started to play up, resulting in a top speed of only 50 mph, then one of the steel engine mountings broke and the fan went through the radiator, making a huge hole. Messages were sent to the MG Car Club asking for help, while Jean and Tom tried, with only moderate success, to circumvent the problem. Eventually a radiator, taken out of another car found on the side of the road, was brought to them and installed and they were on their way again.

The rest of the rally was a mad dash to time controls in an effort to make up lost time, and, though their disasters had cost them points, they pulled into Sydney—the only sports car to finish the event.

It is interesting to see today how women in motor sport appear to have gone full circle. For many reasons, not least the lack of sponsorship and the high cost of motor sports generally, women are now a rare commodity amongst the genre. There are very few female drivers in today's premier championships. Instead, the one place you will see women is on the club circuit, and they are particularly evident in the pre-war cars.

Ann Templeton is a familiar face to many who have seen her racing with verve on tracks up and down the country. She has, since childhood, had an

Ann Templeton at speed during the Silverstone MGCC Meeting in 1991.

interest in motor racing, but it wasn't until she married that this interest was able to develop into anything more. Now she is unstoppable. She gave a J1/J2 special, built and co-owned with Len Bull, its first outing at a Goodwood Sprint meeting in 1990. Ann's determination to race meant that she and the J1/J2 were soon competing in MG Car Club and VSCC events. For her, the attraction of the track was the thrill and exhilaration of dicing with the other competitors, throwing the car into four-wheel drifts or sneaking up the inside of a competitor before powering out of a corner—leaving the opposition scrabbling at her heels.

Winning the Mary Harris Trophy during the MGCC's weekend at Silverstone in 1992 was a highlight in her motor sport career. Ann passed the chequered flag unsure of who had, in fact, won the race. When the results appeared, the winner was someone she had seen broken down at the side of the track during the event, which she thought seemed rather strange. However, once everyone had got their sums right and checked the handicap, she emerged triumphant, proving that even small-engined cars can be competitive.

Despite her success in the J2, Ann added a KN to the stable to satisfy her taste for speed. Her first venture out in the car was to do just that. On a test day at Silverstone she opened up the throttle down the back straight and thought, 'This is very exciting, the speed, the power of the engine, it's just terrific'. Braking where she was used to so doing in the J2, though, had an alarming effect. Going too fast into the corner, she was soon spinning violently. 'It was such an exciting feeling, but I was terrified at the same time. As the car spun round I was scared that we were going to roll, and I did not want to wreck the car on the first lap of our very first time out.' Bringing the car back under control, Ann sped on to finish the practice session and now races this car with alacrity at any opportunity.

Other lady racers of pre-War cars include Elizabeth Green and Carol Cooper; both of whom are regularly seen on the tracks around the country entertaining the spectators and enjoying their motor sport with a considerable measure of success.

Elizabeth Wigg has become an aficionado when it comes to preparing pre-War MGs for Concours d'Elegance. 'It's not worth doing something unless you enjoy it,' she says, and judging by her success she must enjoy it very much.

The 'polishing passion', as it could frivolously be referred to, started in 1985 when Elizabeth bought a beautiful red 1946 TC. It was a remarkably original car but, being an absolute perfectionist, she changed certain details to ensure that everything on the vehicle was completely correct for its age and type. She then embarked on the Concours circus, progressing from one victory to the next and winning the 1988 'T' type championship.

Liz Wigg and Rivers Fletcher with their very successful 1930 Double-12 Midget.

Anna Guerrier driving with her husband, Keith, in the Rally Britannia, 1992.

By this time Elizabeth had decided to tackle an entire restoration job on a J2, not something usually associated with women. Her eye for detail meant great pains were taken and pleasure derived from sorting out any defects, until the car achieved the pristine condition in which it is seen today.

'I take the different parts to the various specialists myself. I find that this is the best way to do it, as I expect extremely high standards. If I deal directly with the professional concerned, there is no misunderstanding. I also like to do as much myself as possible, though on occasions I have asked people to come to our home to help me. I am so grateful for their assistance as I would not have achieved the results without their help and personal attention.'

The awards she has won with this car include Best Car in Show at Knebworth and Best MG in the UK at the NEC MG World. At Silverstone Historic and Beaulieu she took Car of the Show, along with numerous *première classes* and other awards at shows throughout the seasons.

An L2, first registered in April 1933, was purchased in February 1989. Apart from it being a particularly good car, she was very excited at finding out that it had been raced in the heady days before the War. A complete rebuild from the chassis up was embarked upon, in which Elizabeth was the prime mover. As always, every component from split pins to wheel hubs, windscreen to paint finish, had to be attended to with an immaculate and precise eye. The only difference with this project was that the car has been seen competing in all events rediscovering its old racing history under the expert hands of Rivers Fletcher as well as Elizabeth. They also run the 'M' type Double Twelve which won the race of the same name in 1930 and which Rivers drove and tested in his youth at Brooklands.

Sad though it is, the modern woman is rarely seen in modern motor sport while those who courageously venture into single seater, production saloons or rallying are viewed as an unlikely phenomenon.

The year 1992, however, will not only be remembered as the one in which Nigel Mansell won the Formula One World Championship, but also the year that Giovanna Amati, amid much publicity, was signed to the F1 Brabham team. Unhappily, she was nothing more than a pawn in the team's desire to attract sponsorship, which did the increasingly difficult role of women in motor sport no favours whatsoever. Thus the female ranks seem to be disappearing and our motor sport heroes with them. So, I do hope that female readers might feel enthused to climb behind the wheel and have a go, or at least encourage someone they know to emulate the British female racing heritage that we have, before it vanishes into oblivion.

The Immortal 'T' Types

David Saunders

Successful though MG sports cars were on the race tracks of the world in the mid-Thirties, it was not to continue. A management decision taken above Cecil Kimber's head decreed that the marque must 'stop racing and make money'. All the hard-won experience, though, was put to good use and resulted in what have become known as 'The Immortal 'T' Types', launched in 1936 and nowadays perhaps the most collectable of all the mass-market MGs. David, Saunders, who records the history of the various models, has been driving a TC for 23 years and is the Editor of the prestigious *'T' Register Yearbook*. His exhaustive research into the history of the cars and personal contact with the men who made them, such as Henry Stone, George Tuck and John Thornley, give his contribution added authority. He not only puts into perspective the claim that the TC was 'the sports car America loved first', but also offers some fascinating behind-the-scenes information about the development and promotion of the other 'T' types.

The 'T' series MG Midget, announced to the motoring world in June 1936, was conceived in the light of conditions which affected the control of the company owned by William Morris (Lord Nuffield), along with the Morris Garages. Lord Nuffield decided at that time to divest himself of his private interests, which included MG, and so the MG Car Co Ltd became a fully-owned subsidiary of Morris Motors and, therefore, under the wing of the Nuffield Organisation.

The implications of this were that Cecil Kimber was now answerable to Leonard Lord, rather than the more benevolent William Morris. At a lunch in Abingdon early in 1936, Leonard Lord told the MG workforce in no uncertain terms to '. . . stop all this motor racing and concentrate on making money'. The MG factory would no longer be supplied with special engines and equipment, and the company would have to use basic units produced in the Nuffield Group's component plants.

At the time this was viewed as a disaster for the much-loved Midget, for after all, its healthy sales figures were no doubt a reflection of its competition success. However, George Tuck, publicity manager at MG during the Thirties, but now able to look back with a dispassionate view, says, 'I am of the opinion that the old Wolseley overhead camshaft engine was due for replacement anyway and, as it turned out, the pushrod engine was a suitable replacement—and in some ways an improvement.'

Whatever the feelings of the 'passionate' MG enthusiast, the motoring press were quick to latch on to the many positive aspects of a change in policy at MG. The *Autocar*'s announcement of 19 June 1936 was headed, 'The Midget Grows Up', extolling the bigger engine, longer wheelbase and hydraulic brakes (previous Midgets were cable-operated), while the larger cabin space—'. . . and the driver can bring his right arm entirely inside the car'!—along with a generous luggage compartment and 15-gallon fuel tank, made it an excellent touring car.

Regarding the engine itself, this was a four-cylinder unit with pushrod-operated valves, double valve springs and a camshaft driven by duplex roller chain. The crankshaft had three main bearings. An external oil filter was fitted, of the disposable canister long-life type.

The interior was extremely well appointed for a car which, after all, was at the inexpensive end of the sports car market. There were leather seats (one-piece back squab with separate cushions) and door panels with full width map pockets, a fully-trimmed interior in matching leathercloth, pile carpet and walnut veneer dashboard. As for weather equipment, this had been particularly well thought out, with two side curtains per side—both fitted with mica windows—being an unusual feature for a two-seater car of the day, and making the Midget particularly light and pleasant for its occupants when full weather protection was necessary. Equally, the twin bow hood gave the car a most pleasing profile, unlike most sports cars for which weather protection gives the impression of having been very much an afterthought or a concession to the feeble! With nice detailing, such as chrome trim to give a finished appearance to the sidescreens (*neé* side-curtains), and a felt-lined storage compartment inside the car to protect them when not in use, plus a half-tonneau made of leathercloth rather than canvas (presumably to withstand the rigours of regular handling), the Midget really had become a multi-purpose car by 1936.

Even so, the *Autocar* was quick to point out that there was no reason to believe that the Midget had lost any of its sporting characteristics—as the road test would later confirm. To allay such fears, the specification confirming its true pedigree included a fly-off handbrake; Jaeger rev counter

A 'T' type being given a helping hand on the 1939 Land's End Trial. Midgets were a popular choice for this form of motor sport and MG had two trials teams, the 'Three Musketeers' and the 'Cream Crackers', managed by the company's service manager, John Thornley. (Montague Motor Museum Collection)

(mounted in front of the driver) matched to a speedometer on the passenger side; dash-mounted map-reading lamp (passenger side) matched to a 'Thirtylight' for the driver—wired to the speedometer, a green light came on between 20 and 30 mph to act as an advisory speed to maintain in restricted zones (driving with the green light on should avoid the embarrassing 'Good morning, officer!'); a reserve petrol tap on the dashboard that allowed for a safety margin of approximately 100 miles; and a quick-release petrol filler cap on the slab fuel tank—sheer Brooklands-inspired! To complete the sporting package, 19° wire wheels, fold-flat windscreen and large spring-spoke steering wheel ensured that the new 'T' series Midget was every inch an MG that enthusiasts could take to their hearts.

The first road test of the new Midget duly appeared in the *Autocar* of 16 September 1936, nicely timed to give an impression of how it performed 'in anger' prior to being subjected to public scrutiny for the first time at the Olympia Motor Show that autumn. The overall reaction was one that would ensure the MG stand was given high priority on most show visitors' lists—prospective buyer or not—with references such as '. . . in character the car remains of the same type . . . it gives an unusually good performance for its engine size, handles in a distinctly better manner than the ordinary touring vehicle and possesses those touches in the *tout ensemble* that endear it to the owner with sporting tendencies.'

As to the performance that could be expected of the new car, the *Autocar* found that, like the earlier 'P' type Midget it replaced, it was still a car that enjoyed being held at a speed of between 50 and 60 mph, with speeds in the upper 70s attainable. Cornering capabilities and braking were found to be fully competent in dealing with the performance. For the first time on a Midget, synchromesh was employed, although not before the factory had already produced 183 cars without it—something of a mystery, especially as the same type of synchromesh gearbox had been fitted to Morris cars since 1935 and was, by that time, also employed on Wolseley models! Anyway, synchromesh when it did arrive on the Midget provided for smooth gear changing to top and third, an experience which the *Autocar* felt would be scorned by the purists. However, one gets the impression that motoring journalists considered this to be nothing more than a natural progression in the evolution of the automobile. Certainly, the road test report gave no indication that the exhilaration of sports car driving had in any way been devalued by this addition. 'Changing up quickly is greatly facilitated, and rapid, quiet downward changes can be made by speeding up the engine exactly as would be done were there no synchromesh. At lower speeds, the synchromesh engages very well unassisted, and a sure drop can be made to third or second for extra acceleration or a steep gradient.'

The 'T' series MG gives one the impression of having been extremely well thought out by a design and development team who knew their market perfectly—judging by the almost total lack of adverse comment in the motoring press. It has to be remembered that, by 1936, the MG Car Company had a lot of experience and success to its credit, both in competition and in the wide range of cars produced. Even so, there had been occasions when it seemed that the heart had ruled the head in marketing decisions, at Cecil Kimber's express instruction, so all credit to 'the men at Abingdon' for having got their new Midget so right 'straight out of the box'. Small details did need attention, but even these were quickly put right by the factory. The *Autocar* commented that their test car was lacking a sporty note from the exhaust, earlier Midgets having been loved for their distinctive exhaust burble. However, the required 'MG note' was quickly sorted out—even to the extent of modifying the test car for the *Autocar* to try again! Such was the importance given by Abingdon to ensuring that their new Midget had enthusiast appeal in every department.

The purpose of this chapter is not to provide the reader with a work of reference on 'T' type specification changes, or to delve into matters of originality. There are many specialist MG books available to assist the owners and restorers of these cars, books even that are concerned purely with the 'T' series Midgets; indeed, the subject is so vast that it would be more difficult to know what to include in the limited space available here than what to leave out. So, it is my intention to outline for you only the gradual progression from the first 'T' series model in 1936 to the last one off the production line in 1955—the longest production run of any one designated MG model (including the MGB). One must not forget, though, that the Second World War interrupted car making at Abingdon for a five-year period, and the 'T' type did go through three body changes in that time, so maybe my statement would prove to be a contentious issue amongst MGB enthusiasts!

To the untrained eye and the lay person, the 'T' series Midget might appear to have continued unchanged up until the outbreak of war, and beyond. In fact, there were many changes that the MG enthusiast was aware of which made the 1949 Midget—the last year in which the same body style was produced—a very different animal from that conceived in 1936.

It was during the early days of 'T' type production that the MG Car Company dropped its policy of announcing an annual programme for each model, and instead adopted a plan of introducing improvements and changes in specification whenever felt desirable. Therefore, to the restorer of one of these cars for whom originality is all-important, sorting out just what is and what is not correct on a particular car can become an absolute nightmare. Add to this the fact that, for the sales department at Abingdon, customer satisfaction with the personal touch was most important, to the extent that special order requirements would be catered for wherever possible, and one will begin to understand the veritable minefield that the poor *concours* judge in a rally field of 'T' type cars today is confronted with! Matters might have been helped had not the factory records for these cars been destroyed many years ago—a moment's lack of foresight on the part of someone in the administration block who was possessed by the 'springcleaning' bug. Today one must rely on the dedication of 'the Few'—those prepared to seek out knowledge and impart it to others, either through a book on the subject (of which there are many) or by way of the MG clubs that now flourish throughout the world. One word of warning, however. Never look upon a restorer's guide as being of Biblical accuracy, if you get my meaning. In the final analysis you, the owner, must decide what is right for your car—for better or worse!

The first modifications to the 'T' series MG came in 1937, when even the untrained eye could not help but notice that the car was looking rather different from behind. In fact, the rear wings had been widened and strengthened to give an altogether more substantial appearance, and from then on all 'T' types would have a distinctive rib running down the centre of the wing. To accommodate this, the slab fuel tank had to be reduced in size to

13½ gallons. Aesthetically, this was a good decision, as the proportion of tank to wing looked decidedly wrong on the early cars. At the same time, side-laced wire wheels were replaced by centre-laced ones. Only the eagle-eyed would have noticed that improved ventilation to the engine compartment had been provided by increasing the number of louvres in the bonnet sides from 15 to 21.

At about the same time as these first changes took place, the *Motor* had the opportunity to test the new car's performance at Brooklands. Even in those days, when the famous circuit was still in use, the *Motor*'s journalist commented on the large cavities in the track that prevented safe driving at full speed. Even so, the following results were obtained. 'By a checked and accurate speedometer it was found that the road maximum was up to 77 mph with the screen up. This was a car that had done only 1,000 miles, and we believe that on further running with the screen down the 80 mph mark should be exceeded, as it is on a staff car that has done some 8,000 miles.'

That such speed could be achieved from a new engine was all the more creditable given that the mixture was set up for economy running, resulting in an average of 33 mph over some 300 miles of hard driving with full use of the gearbox.

In talking of speed, many people instinctively related inherent danger, with sports cars encouraging such practice. So suggested the *Motor* in the course of this particular road test, in order to emphasize the safety margin built into the Midget from its competition pedigree. One experience in particular meant that controllability had to be tested in earnest. During the course of lapping the Brooklands circuit, an 'excavation' about 2 ft across and 1 ft deep suddenly appeared across the bow of the car while travelling at 70 mph.

'By hard braking, and putting the car into an intentional slide, disaster was avoided with real thankfulness for rapid response in time of need. Such contingency directs one's mind very forcibly to a merit of this type of car that is not generally appreciated—that is, the high margin of safety. "Safety Fast" is, of course, an MG motto and one that is well deserved. It is stated in some quarters that sports cars must be dangerous because insurance companies demand a higher rate of premium on them. This is entirely false reasoning. The consequences of an accident at high speed in general are more severe than one at low speed. Thus, fast drivers as a rule "cost" more than slow drivers, even though they may not be prone to more accidents numerically. In fact they will be more prone if they indulge in fast driving in unsuitable cars, and the fact must be faced that there are cars which will achieve high speeds, but lack full control when so used. On the other hand, the MG remains safe and stable up to its maximum.' So wrote the *Motor* in December 1937.

Reading such reports of the period can often be most enlightening, as the car that confronts you today has had a lifespan far beyond that ever envisaged by the MG factory. In all probability it will, at some time in its life, have been subjected to misuse, inadequate servicing, misconceived modifications and inferior replacement parts and, even now, despite an MoT certificate, contain worn parts that need renewing. Fortunately, the MG enthusiast is well served, with many excellent specialist suppliers stocking almost everything you could possibly need, so there is no reason why a 'T' type should not be as enjoyable to drive today as when it was new. However, such glowing reports on the part of the *Autocar* and the *Motor* should be considered as nothing more than a yardstick as to what is attainable, not necessarily what your prospective purchase will drive like when you first get behind the wheel.

Having dealt with the matter of individual idiosyncrasies, I feel on safer ground to relate other qualities that appealed to the *Motor* on its first acquaintance with the 'T' series Midget.

'It must not, however, be thought that the sole attractiveness of this car is in such utilitarian matters as speed, mpg and safety. It is the fortunate possessor of a really attractive personality. When driving, one becomes part of the motor car and experiences once again real joy in driving, a sensation apt to be forgotten when one employs the mere transport of a conventional saloon.'

In referring to the weather equipment as being something of a 'paradox', the Midget being '. . . in every way equal to that of the average closed saloon', and with metal-framed sidescreens that '. . . convert it into a coupé in all but name', you may begin to think that the *Motor* was perhaps getting a little carried away with its enthusiasm for the car. Maybe, but then you have to

A 1939 Tickford totally restored by Bob Douglas in 1989, 50 years after its manufacture.

bear in mind that, prior to the 'T' type, sports cars were not noted for their blessings during inclement weather.

The Tickford

However, if the standard Abingdon product fell somewhat short of the mark in standing a direct comparison with saloon car comfort, another Midget was just around the corner which would certainly justify the analogy. In August 1938 a drophead coupé on the 'T' series chassis was announced, built on the Tickford principles by Salmons and Sons Ltd of Newport Pagnell. The car, Abingdon-constructed forward of the scuttle, was then driven the 50 miles to the Salmon factory, where a most luxurious two-seater body was built on to the chassis. To facilitate a draughtproof environment for the two occupants, free of wind noise and rattles, a lined hood of solid construction with 'pram-like' toggle irons fastened to fixed windscreen pillars, but with the windscreen allowed to open out and upwards to a horizontal position ('a valuable asset if one is driving in fog' said the *Autocar*!). Shoulder-height doors were fitted with wind-up glass windows. Other features included scuttle-mounted windscreen wipers, and semaphore indicators recessed into the body forward of the doors (the standard 'T' type was not fitted with direction indicators).

Another 1939 Tickford restoration—the view from the driver's seat of American Bill Hentzen's car.

The meaning of luxury in full leather in the front seats of the Tickford MGs.

Inside, the car was pure luxury—individual leather seats, deep-pile fully carpeted floor with edge binding, walnut capping on the doors, telescopic steering wheel adjustment, an interior light and a dash-mounted ashtray. The price of all this luxury? Just £47.50 more than the standard Midget, £269.50. Mind you, in today's world of inflated prices, such an apparently modest increase is difficult to measure.

At the same time as the Tickford Midget was announced, Abingdon also introduced a wide range of new colour schemes for the 'T' series car. Previously this had been limited to saxe blue, racing green and carmine red with interior trim to tone, plus black coachwork with blue, green or red upholstery and trim. (Note: the wheels were always painted silver as standard). From the late summer of 1938, however, not only was there a new range of colour schemes but, for the first time, the option of two-tone colour combinations and also a metallic paint. The colours available were black, saxe blue, coral red, apple green, duo green, maroon and metallic grey, while the leather upholstery and leathercloth trim was in a choice of blue, red, green, brown, maroon, grey and biscuit. For the standard Midget, weather equipment was

107

only supplied in black, but on the Tickford model maroon, blue, grey and fawn hooding was also available.

Any combination of the above colours could be selected at no additional cost. For two-tone coachwork, the body was finished in one colour while the wings and fairings (front valance and running boards) were in the other. Metallic grey was only applied to the body, the remainder being painted to match in a non-metallic lustre, the reason being that the areas of the car prone to damage could then be easily repainted.

The TB

In the following May a new 'T' series Midget appeared, designated the Series TB. It was intended that this would spearhead the MG programme at the start of a new decade—the Forties. However, in line with Abingdon practice, production was not held back until the proposed date of the Olympia Motor Show that autumn; indeed, its announcement was not even heralded in a blaze of publicity. A review of the new car did not appear in the *Motor* and the *Autocar* until early September—after the official declaration of war and by which time Abingdon was already shutting down its production lines to make way for a contribution to the war effort. The TB Midget had come very close to being stillborn. As it was, just 378 cars rolled off the assembly line between May and September in that long hot summer of 1939.

As a consequence of the circumstances that were to affect the world for the next five years, the TB Midget was to become the rarest of all the 'T' series cars. Externally there was no difference—not a single clue. All the changes were to be found under the bonnet. Little wonder that the motoring press saw no need to make a drama out of the car's launch. After all, the average motorist was hardly likely to be inspired by a different engine. In fact, the development was one of the most significant to happen on the technical side throughout the 19-year history of the 'T' series Midget, as the new power unit would continue to be used up until the end of 'T' type production in 1955, and also be the source for a great deal of competition success in racing specials after the War.

For the technically minded, the new XPAG engine, as it was designated, had four cylinders of 66.5 × 90 mm (1250 cc) whilst the earlier car, which from now on would be referred to as the TA model, had four cylinders of 63.5 × 102 mm (1292 cc). Thus it can be seen that the improved engine of the TB had a larger bore, a shorter stroke and slightly smaller capacity. The peak rpm was therefore raised from 4,800 to 5,252 to compensate for the slight reduction in capacity while, to ensure reliability at continuous high operating speeds, the crankshaft was counterbalanced and the bearings of the thin-wall type fed with oil at high pressure. What this all meant was a livelier performance for the new 'T' type—better acceleration at the expense

In 1945, as there became less war work for the MG factory, there was time to think about car production again and here the *Autocar* road test TB provides the basis for the prototype TC. (MG Car Co.)

of a low-revving engine at touring speed. In the opinion of Henry Stone, a development engineer at Abingdon for over 40 years, 'The XPAG unit was a great improvement on the one fitted to the TA, which was very much overweight, and with a heavy oil/cork clutch assembly as well.'

The other major change with the TB Midget was the installation of a Borg & Beck dry plate clutch to replace a cork insert clutch running in oil, and the gearbox now had synchromesh on second as well as third and top. Apart from these changes, the TB was identical in every respect to the earlier TA, and no further modifications to the 'T' series Midget would be made until production restarted after the war, re-emerging as the 'famous' TC—and with more important specification changes.

The TC

Production of the TC Midget began in November 1945, a car of identical appearance and similar proportions, but available only in restricted numbers to the MG faithful in the UK anxious to put the misery of the past five years to one side. Unfortunately, the period of hostilities was to be followed by an equally long period of austerity, which meant that only a small percentage of Abingdon's total production would be available to the home market. There was a shortage of steel for the building of cars after the War because so much was required for the reconstruction of housing and industrial plant which had been damaged by bombing, so allocations were based on a company's export record. For the Nuffield Organization, which had exported a growing proportion of its cars in the late Thirties (but not necessarily MGs), this meant there was no problem, so Abingdon had to be extremely thankful that it was under the Nuffield umbrella!

It so happened that, in the TC, the MG Car Company found itself with an extremely marketable commodity abroad. While never produced in left-hand drive form, most exports at that time were to far-flung regions of the British Empire where centuries of colonial influence ensured that everybody resolutely drove on the left-hand side of the road. However, it appears that the delights of MG motoring had captured the hearts of American servicemen stationed in England during the War, as it was not long before Abingdon began to receive orders for their product, albeit in right-hand drive form, from across the Atlantic. At first it was just a trickle, with only a handful of cars exported to the States in 1947, but during 1948 the demand was such that this could only be satisfied at the expense of customers at home. Indeed, such was the impact that this little car made in a relatively short space of time that, as thoughts turned towards a replacement for the TC, a left-hand drive option would become part of the Midget's specification for the first time. Whether it was the TC or its successor that paved the way for a whole new market for British sports cars depends very much on your interpretation of events. Whatever your point of view, it was Abingdon that 'cast the die'— a role that the British car industry would continue to play throughout the Fifties, Sixties and Seventies; indeed, for as long as there was a mass-produced British sports car worthy of the name to sell abroad.

Amongst British enthusiasts, for whom supplies of the MG TC had been limited, this car is often presented as the archetypal British sports car. So why, when it was simply a progression of a pre-War design and already outdated in technical specification when it first appeared, should it be held in such high esteem? First, it has to be borne in mind that only 3,000 TAs were produced, while of TBs there were only a few hundred. The TC, however, reached a production total of 10,000, which meant that, even allowing for the export quota, there were still just as many TCs available for the home market over its entire production run as 'T' types produced before the war. Furthermore, it had the superior engine introduced on the TB—and the weaknesses of the XPJG power unit fitted to the TA were to come to light more and more with age. As to why the TC should also achieve greater

Export of TCs to America began on a large scale in 1948. Here a team of drivers—exposed to the elements!—are on their way with one of the first consignments to Liverpool docks. (Roby Cruyswegs)

A TC bound for America being loaded into the cargo hold of a ship at Liverpool. (Roby Cruyswegs)

popularity with MG enthusiasts in Britain than either of its successors, the TD and TF, this is simply because it represents the last of the truly traditional British sports cars, not so much in appearance as in its technical composition. It was the last beam-axle MG; its lighter body gave it a considerable competitive edge over later Midgets, while its large diameter wire wheels and rudimentary steering box made it instantly responsive in the hands of the experienced driver. With the start of the Fifties—and, in the eyes of many, the dawn of the modern age—comfort was understandably to play a more important role in what was on offer to the sports car enthusiast.

So exactly what was it that Abingdon had to offer in 1945? Well, for one thing the TC Midget was the only MG model available. Gone for ever were the larger saloon and tourer models that had been produced alongside the Midget before the War—but then I suppose we must be thankful that the MG factory survived without bomb damage, unlike the Cowley engine plant which received several direct hits. Rationalization now had to go hand in hand with the austerity measures, so this meant just one basic model for the time being—not even the Tickford variant of the Midget had survived. However, changes for the better in the basic Midget product there most certainly were.

It is reasonable to believe that, throughout the five years when the Abingdon factory was concentrating on the war effort, men whose whole lives had previously been devoted to building cars must have given the occasional thought to how their Midget could be further improved. Certainly, with the end of hostilities in sight, there was more time in which to give some attention to a prototype. One of these men was Henry Stone who, in a letter to me explaining why the problems of fuel vapourization with a hot XPAG engine had never been addressed by the factory, wrote, 'At that time (1945) we had no development shop at Abingdon, thus no opportunity for extensive testing. It was necessary for survival that a Midget be produced again quickly, so the assessment of the TB's technical failings for lack of experience was mostly confined to the suspension.'

So it was that, only six months after the end of the Second World War in Europe, the first TC Midgets were being rolled off the production line. The changes involved a wider body (an extra 4° between the rear door pillars providing more seating room) and rubber-bushed spring shackles (the leaf springs had previously been mounted in sliding trunnions that required regular greasing). The battery, previously two 6-volt units carried on the chassis beneath the luggage compartment, now became a single 12-volt unit housed as an integral part of the bulkhead. One further modification was to the dashboard, where a petrol warning light replaced the petrol reserve tap fitted to pre-war Midgets—a green light displaying the word FUEL started flashing when the petrol tank contained only 2½ gallons and, from about the 1½-gallon point, remained on constantly.

Colour schemes reflected the general state of the world at the time—'any colour as long as it's black' (to borrow a phrase first used by Henry Ford many years before!). At least there was a choice for upholstery and trim; red, green or beige. The colour of the interior also appeared on the radiator grille, which helped to break up the drabness a little, as with the weather equipment which appeared for the first time in fawn canvas instead of black. However, with materials in general being in short supply, strict adherence to specification was not always possible, and it is therefore highly likely that early TCs left the factory fitted with black hoods and sidescreens in order to keep the production line on the move. Certainly, the black leathercloth half-tonneau remained as standard well into 1948.

During the lifespan of the TC Midget numerous detail changes were made to the specification, most of which can be related fairly accurately to chassis number in specialist books on the subject. Owners and restorers of these cars should therefore have little trouble in ascertaining what is correct on a particular car—that is if, like most, you eventually get bitten by the 'originality bug'! Despite such austere beginnings, the many specification changes that took place during its four-year production run make the TC one of the most interesting models to restore, as the variations help to make each car a little bit individual. For example, with regard to colour schemes, it was less than a year (June 1946) before Regency red and shires green were available for the coachwork, and by 1948 sequoia cream and clipper blue had extended the choice still further, although upholstery remained restricted to just the three shades. There was, however, a number of options on contrasting coachwork/trim colour combinations.

Examples of the many other specification changes that took place during the production run of the TC—some forced on the factory as parts were redesigned by the accessory suppliers, others more the whim of Abingdon itself—include the headlamp lens (flat to convex glass with different pattern fluting); foglamp (late TCs had sealed reflector unit); dashboard (walnut veneer replaced by leathercloth covering to match trim); hood (twin rear windows changed to a single panel); engine colour (light grey, then grey-green, finally maroon); bulkheads (engine colour on early cars/coachwork colour with maroon engine).

Despite such a multitude of modifications, there are distinguishing features that make the TC instantly recognizable and tell you that this car is not a pre-War 'T' type. The most obvious of these are the running boards, which contain only two tread strips instead of three, to allow for the wider body. For the same reason, an angle had to be introduced into the profile of the sidescreen frames in order that they would still fit snugly under the hood valance, although it takes a keen eye to notice the difference. Interestingly,

113

this practice was continued on to the TD and TF models. More noticeable was the D-shaped rear lamp, whereas the TA and TB had a circular lamp. However, this detail can no longer be relied upon for identifying the car, as when new lighting regulations were introduced at the start of the 'Fifties the original lamp became illegal and owners had various options open to them. Some fitted a pair of D-shaped lamps of a new pattern in the correct position, either side of the number plate, while others fitted a standard type of new tail-light to the rear wings.

The TC Midget was used quite extensively as police patrol cars (as indeed was the TA), and should you be at all uncertain as to the origins of your particular car, such noteworthy beginnings will be confirmed by a larger 'blister' on the nearside bonnet panel to accommodate a bigger dynamo. This was required to compensate for a larger battery which, in turn, was required to power the additional electrical components—two-way radio, loudspeaker, etc. To accommodate the heavy duty battery, a modification had to be made to the bulkhead, allowing for a wider battery compartment.

Well over half the entire TC production took place during 1948 and 1949 while, as late as December 1948, with less than a year of production still to go, Abingdon produced a special version for the American market. While retaining right-hand drive (no doubt because a totally new Midget was in the planning stage and the technical implications made the changeover impractical) modifications were made, mostly to satisfy US traffic regulations. Noticeable was the inclusion of bumpers at the front and rear, smaller head-lights and a different rear lamp arrangement. There were many detail changes but the most interesting was the inclusion of turn signals in the form of flashing indicators (twin filament bulbs in the front sidelights, rear lamps mounted high up on the petrol tank sides). At this time British cars were still supplied with semaphore indicators—although the TC, like its prede-cessors, was devoid of any form of signalling—and flashing indicators were widely regarded by motorists in the UK as simply an American gimmick that would never catch on!

Exporting cars on a large scale was a new experience for the MG Car Company, and in the early days there were lessons to be learnt. John Thornley, sales and service manager at the time, later becoming general manager and then a director of the company in 1956 (and now President of the MG Car Club), recalls how the cars would be mistreated during shipping. 'With early shipments we didn't realise how dreadfully the cars would be knocked about during handling at both ends of the sea voyage. On one occasion a docker was discovered walking along the bonnets of adjacent cars packed side by side in the hold!'

However, Abingdon soon mastered the art of exporting cars. 'We hadn't been shipping very long before the cars were stripped beforehand of everything readily removable and fitted with various forms of armour—e.g. baulks of timber in place of bumpers—before being released to the tender mercies of dockland!' Eventually, cars would be despatched in 'knocked-down' form, packed in strong wooden crates. As John Thornley recounts, though, even this method was not foolproof, as in the instance of a consignment to Australia. 'When the case was opened in Sydney, there inside were found a cat and four kittens, all alive and well. The packing crate had been screwed down at least five weeks earlier in Cowley. Every spot of lanolin, with which all brights parts had been liberally coated, had been licked off! I gather that the stowaways were made a great fuss of at the time.'

The TD

After the depression of the Forties, in Britain everyone was looking forward with hope and optimism to the start of a new decade. The Festival of Britain exhibition on the South Bank in 1951 would echo these thoughts, but Abingdon had already made its own contribution even before the bells rang in the first day of 1950. This was a brand new Midget sports car, the TD, and the first 100 had already been built by then although, as neither the *Motor* or the *Autocar* announced the new car until January, it is highly unlikely that any were registered before that date other than road test cars. Indeed, the *Motor* was ready to recount its experiences of the first 900 miles in one of the

A superb 1953 TD Midget with red weather equipment fitted by the owner, Rod Sawyer.

new MG Midgets as early as 22 February, 1950.

The traditional styling of the new Midget was received with mixed bless-ings. The *Autocar* began its introduction to the new car thus: 'Meet the new "Midge". Everybody calls the well-loved MG Midget the Midge, so it might just as well be printed. The Series TD may be new, but it still looks like a Midge, and has not "gone all futuristic", for which many thanks, people will say. A sports car ought to *look* like a sports car, and its innards ought to be accessible so that fans can personally keep it in tune; they should not be hidden beneath billows of bent tin.' However, it was widely felt at the time that something just a little bit more exciting and in tune with the modern age had been hoped for.

As a result of the success that MG had achieved in the States with its TC model, other British car manufacturers were also now looking towards this potentially lucrative new market on the other side of the Atlantic with a product of their own, and, unlike MG, they had no allegiance whatsoever to traditional sports car values. Before long the Jaguar XK120, the Austin-Healey 100 and the Triumph TR2 would all be competing for the same market. Had Abingdon misjudged the mood of the time?

Maybe it was just that the men at Abingdon found it difficult to shake off a lifetime of tradition, or maybe they felt that there was too much at stake to risk such a giant step. More likely I suspect that their sales figures with the TC had told them that there was a ready market abroad for a British sports car built on traditional lines and with traditional styling—and it was exports that mattered most.

If that is so, then the strategy certainly paid off, as the TD Midget did prove to be the marketing success that was hoped for. In just three short years of production, almost 30,000 cars were sold, of which 28,000 were exported. Less than 2,000 were sold on the home market, a fact underlined when one considers that this was under half the number of TCs sold to UK customers. However, whatever the sports car enthusiasts in this country may have thought of the TD, the MG Car Company had made its mark as a major exporter (a position it would continue to hold right up to the end of its Abingdon days) and the MG name became synonymous with sports car enthusiasts the world over—but especially in America, where the vast major-ity were sold.

So just *why* did the TD get such a poor reception at home? The answer is to be found in the very reason for its export success—traditional shape with-out the traditional discomfort! In the mind of many British enthusiasts at the time, the TD was pretending to be something it wasn't, while to the average American hooked on anything English, the addition of a few 'home comforts' was not such a bad thing. Hence the independent front suspension, rack and pinion steering, easy-clean wheels, bumpers with overriders, and a dashboard glove compartment that allowed space for a radio, turned the MG Midget from a plaything into a practical road car.

Today, of course, we live in a very different world. Forty years on from when the TD was last in production, it is now seen as a very desirable MG sports car in the traditional mould for those who do not have masochistic tendencies—and an excellent touring car providing all the fresh air motoring that two people could possibly ask for. It may not win any prizes in the competition stakes, its body being a little on the heavy side for the 1250 cc XPAG power unit installed, but, for all that, its practicality and comparative 'civility' make it a firm favourite amongst MG enthusiasts today.

The TF

If the TD was considered a disappointment amongst ardent MG enthusiasts back in the Fifties, that was nothing compared with the birth pains that TF had to endure, for here was a car that even Abingdon did not want to build! Today, ironically, the TF is considered by some to be the most desirable of all the 'T' series models, especially in 1500 cc form, the XPEG power unit providing a much-needed sparkle to the performance in the last six months of its production run.

To explain the reasons for the car's unpopularity at the time, it is necessary to understand BMC politics. The Nuffield Organisation had become the British Motor Corporation early on in the Fifties when two car giants, Morris and Austin, joined forces. The management then found itself with two sports car marques in direct opposition to one another—MG and Austin-Healey. Leonard Lord, you will recall, had been instrumental in the concept of the 'T' type MG, but the working relationship with William Morris was a tempestuous affair and it was not long before Lord had joined forces with Herbert Austin. Now, however, as head of BMC, he found himself back in charge of the Morris empire again. At this time MG had already got an experimental streamlined sports car in its development shop; indeed, they had even gone motor racing again with it! This was a special-bodied MG TD which had put in an appearance in the 1951 Le Mans 24-Hours race. Eventually, a similarly styled car would become the MGA, but Leonard Lord was understandably reticent about allowing it on to the market, in case it harmed the sales of his own brainchild, the Austin-Healey sports car, which had been developed out of a marketing agreement between Austin and Donald Healey for a car to rival the MG! It was into this climate that the TF Midget was born.

Like the TD before it, the TF looked a totally new car, but, in fact, mechanically it was based on the TD and only the coachwork and interior were revised. In many ways it is an extremely pretty motor car, but maybe a design that holds more aesthetic appeal for the trained eye today than when

117

it first appeared in 1953. Between October of that year and May 1955, a total of just 9,600 cars were built, of which 8,000 went for export—once again, the vast majority to the United States. Instantly recognizable by its sloping radiator grille, faired-in headlamps and individual seats (plus wire wheels as an option) the TF has built up something of a cult following in recent years, to the extent that modern reproduction examples can command as high a price as the real thing! Maybe it was a car built before its time. Whatever your own personal view, it brought to an end an illustrious chapter in MG history.

The 'T' series Midgets are the link between the motoring past—the era of Brooklands, riding mechanics and aero screens—and the modern age, when all-steel construction, fully-enveloping bodies and aerodynamic styling would be uppermost in the car designer's mind. During that 19-year period the Midget had grown from being a car purely for the enthusiast, with few creature comforts (as any 'P' type owner would no doubt testify!), intended more for fun than practicality, into a vehicle of adequate proportions suitable for everyday use and with appeal to a wide-ranging age group. In the course of its gradual development, the 'T' series MG no doubt lost a few friends along the way, although this will happen in any age as the 'purists' attempt to hang on to the past. However, the 'T' type Midget undoubtedly gained many more MG enthusiasts than it lost. It also held appeal for many famous personalities back in the days when it was new—household names too numerous to mention here, such as the Duke of Edinburgh, who courted the future Queen Elizabeth in a TC; Graham Hill, double world champion, just one of many famous racing drivers who 'cut their teeth' behind the wheel of a 'T' type; and Richard Burton, the actor, who regarded the TF as his favourite car of all those he had owned. For some recruits the Octagon badge was to become a religion, a drug, a way of life. Regardless of that, marketing success has to be measured in terms of sales—and in that respect the 'T' series cars, as originally conceived, were most certainly the saving grace for the Abingdon factory. In one aspect, though, the 'T' type remained faithful to its pedigree. To the end it was always purely a two-seater.

History lives on! The proud drivers of a group of police MG Midgets pose with their vehicles in 1947. Vehicle FBT 112, in the centre, is now equally proudly owned and maintained by enthusiast Chris Alderson *(see below)*.

My thanks to Chris Alderson and Rod Sawyer,
'T' Register Registration Secretaries of the MG Car Club,
for their assistance in verification of the text.

MGB: 'The World's Most Popular Sports Car'

Geoff Shirt

Thirty years ago, in 1962, MG introduced the sports car which redoubled its fame in the Sixties and Seventies. Indeed, the reputation of the MGB even survived the closure of the company in 1980 and has now provided the inspiration for the rebirth of the marque in the form of the new MG RV8. There is, in fact, probably no other model of British-made sports car still seen in such numbers on the roads of the country. Geoff Shirt, who has written this latest (instead of final, as it might so easily have been!) chapter in the MG story, is one of the busiest men in MG circles, both as a writer and as the organiser of special events such as the London to Edinburgh run and the annual Kimber run each April to celebrate the birthday of the creator of the legend. His perceptive review of the 'Adder'—as the newest addition to the MG range was code-named—brings the story right up to date . . .

The MGB has accurately been described as 'the World's most popular sports car', with over a half-million examples built between its introduction in 1962 and the closure of the Abingdon factory in 1980. Although undeniably true, such a statement belies the car's real strength, the versatility of the MGB's design to be adapted to the changing needs and fashions of the buying public. It is the car's ability to change, whilst not alienating the traditional enthusiast, that has made the MGB so special, so close to people's hearts. On casual inspection, the car could be likened to the parody of the 'original' hammer that had only received two new heads and three replacement handles, even the later emission-conscious model bearing little more than a passing resemblance to the MGB that glistened with chrome some two decades earlier. In

120

The unmistakable lines of an MGB—although production finished in 1980, hundreds of the models were not registered until 1981. (The RoadRunner Collection)

truth, the evolution of the 'species' retained a higher than average number of components over its productive life, albeit through lack of investment.

Indeed, the concept of the MGB drew heavily on two previous MG models built at the Abingdon factory, the 'Z' type Magnette which entered service in 1953 and the MGA two years later. The former had been introduced as a corporate 'sports saloon', the latter as a 'revolutionary' replacement to the long line of wooden-framed 'square-riggers'.

'Z' Type Magnette

Although bearing the same name as the pre-war racing MGs that ruled almost all the European circuit, the 'Z' type Magnette was a rather staid, four-seater family saloon. It was the brainchild of Gerald Palmer who, having designed the Javelin for the Jowett Motor Company, joined the Nuffield Group in 1949. His joining brief was to design for William Morris a new Riley, MG and Wolseley saloon. To make optimum use of a given space, he designed a body that incorporated the axle mounts, technically described as a monocoque construction. This eliminated the deep chassis rails that ran the whole length of contemporary motorcars, and the passenger cabin was made both very spacious and given a lower centre of gravity. Using a further technique, now described as 'badge-engineering', Palmer took this basic unit and fashioned three similar yet individual cars, namely the Riley Pathfinder, Wolseley 4/44 and the MG 'Z' type, using various combinations of unavoidable and deliberate changes. Although constructed as a saloon, the structural strength was in the sills below the door, allowing the same constructional principle to be applied to convertibles.

121

A fixed-head MGA attracts admiration from MG enthusiasts and members of the public alike. (The RoadRunner Collection)

The MGA

During the years leading up to the War, construction of motor car bodies had invariably followed that of the carriage; metal skinned panels over a wooden frame, wings both separate and swept, radiators held erect and running boards providing a decorative and purposeful role. Although continuing for several years afterwards, technical advancement, and indeed fashion, had brought the Motor Industry by 1952 to one of the greatest watersheds in its history, the all-enveloping body. Donald Healey was an early exponent of this innovation and, following an agreement between himself and BMC's managing director at the Motor Show that year, founded a new marque, Austin-Healey, to take a flagship role in the main export market. MG sports cars had previously sold extremely well in the United States but were losing market share, chronically starved of the investment needed to replace the archaic 'T' series Midget. By contrast, a new assembly line was set up for the Austin-Healey 100/4 at the Longbridge plant in Birmingham from 1953. However, pitted against the combined strengths of the Triumph TR2 and the Jaguar XK120, the 100/4 suffered two years of poor sales figures as BMC's American franchises were continually asked, 'Longbridge? Where or what is Longbridge?' Time and again came the call for a new MG from Abingdon. In 1955, prayers were answered and John Thornley was allowed to reopen the Abingdon drawing office.

Ironically and thankfully, the design and prototype for an MG with an all-enveloping body existed; the record-breaker based on a sleek 'TD' that raced at Le Mans in 1951, pre-dating even the formation of BMC. Wherever possible, existing tooling from within the group was to be used and so Sid Enever, who began the project some four years earlier, drew extensively from the Austin, Morris and Wolseley parts bins for mechanical components. Although Palmer's pioneering work with the monocoque bodyshell was at this time available to Nuffield, time was of the essence and the production line assembled roadster and fixed-head bodies based on EX 175 on separate rolling chassis, ready in months. At no time during its first four years could production of the MGA keep pace with demand, even though Abingdon's capacity was increased almost twofold. Approximately 90 per cent were exported to America, a trend that was to continue for many years, with over 100,000 units built in all. As a further paradox, Austin-Healey production was moved to an adjacent line in what the Americans referred to as 'the Sports Car Factory'—Abingdon.

The Concept and Planning of the MGB
Today, as then, work on a future motor car begins almost as soon as a project dream becomes production reality. Hardly surprisingly, therefore, that Enever had already begun to consider a more streamlined version of the MGA in 1956 when Don Hayter and Roy Brocklehurst joined the team. Hayter, having spent the previous three years in the drawing offices of Aston Martin, had been at the other end of the 'Z' type Magnette story; converting

A 1962 MGB faithfully restored to its original specifications of tartan red coachwork, matching hood and whitewall tyres. (The RoadRunner Collection)

Palmer's designs into factory tooling at Pressed Steel! Extremely conscious of the world about him, Hayter was very aware of the shortcomings of the model that had only just begun. Although it had been a dramatic departure from the 'T' series in terms of styling, the MGA had entered service as a four-year-old design with narrow cockpit, basic interior, rattling side-screens and a rear compartment more resembling a luggage rack than a boot! As plans for the spartan MG Midget were well advanced and ready to be announced in 1961, the design team aimed the MGB at a more 'up-market' clientele who wished to take the car for the occasional touring holiday as well as putting it to mildly sporting use. Equipping it with such 'luxuries' as wind-up windows, exterior and locking door handles, some stowage room behind the seats (for young children), it was given a boot capable of holding at least two medium-sized suitcases. With no obvious alternative power unit to the MGA's engine, the 'B' series block was enlarged to at least retain performance under the increased expected weight. The advantages of designing a monocoque construction were clear, the new car having obvious pluses of space and handling, not simply over its predecessor but all its rivals as well. Just as the smooth-fronted MGA had retained the 'spirit' of the distinguishable 'mirror-image' MG grille, the new model took the concept a stage further using a very shallow grille made up of vertical lathes. Although initial drawings had been completed by June 1958, it was exactly four years before the MGB went into production.

Announced at the 1962 Motor Show, the MGB was greeted enthusiastically by the public and motoring Press alike. Over its predecessor, it had a higher top speed, better acceleration and even a marginal improvement on fuel economy, and when compared with its rivals, the MGB gave excellent value for money at £950. As a bonus, it opened up a whole new social class to real 'sports car performance'. On the first test session for *Motor* magazine of an ex-delivery standard car, they recorded 118 mph! Available in 13 different coachwork finishes, many with matching hoods, the car quickly outsold everything in its class. Whilst the bodyshell was brand new, proven mechanical components ensured it would be an extremely reliable and well-tested vehicle from the outset. Due in part to its comfortable leather seats, softer suspension and forgiving handling, most criticism came from 'die-hards' MG enthusiasts themselves. Many felt the refinements had gone too far ... 'Wind-up windows ... optional heater ... on a sports car? They'll be putting carpets in next!' For the record, black carpets replaced rubber mats as standard equipment in 1976!

Development of the Early MGB
The MGB's shell had been an incredibly complex and expensive structure to manufacture. John Thornley was to comment some years later that he

John Thornley, the man who oversaw the production of the MGB, photographed recently with his 1969 MGC Roadster. (The RoadRunner Collection)

realised it would be MG's last new sports car when he received the tooling bill from Pressed Steel. But owing to the long period the MGB spent in the development stage, and the liberal use of existing mechanical components, only one significant modification appeared during the first five years of production—a weakness brought about by increasing the engine's internal size almost 15 per cent without preparing the crankshaft for the extra torque. Once the simple modification had been made, crankshaft failure gave no further trouble, even when the engine was bored out to the limit and prepared to the highest level of competition tune. It was during this stage in the car's life that its suitability for competition came to the fore, albeit of a somewhat restricted nature. Campaigns to race circuits such as Le Mans, Nurburgring and Sebring all brought a level of success, but its potential as a rally car is only now becoming fully appreciated. Sharing 'Comps' with the big Healey and diminutive Mini overshadowed its very able capabilities in this field.

Introduction of the MGB GT in 1965 was important not simply because it gave a model more suited to Britain's inclement weather, but because it opened the marque up to an entirely new and enormous sector of the motoring public that would not have dreamt of buying an open car. From the introduction of the MGA, John Thornley had harboured a desire to produce an affordable fastback, a 'poor man's Aston Martin'. With one of their ex-employees now installed in the drawing office and the most popular roadster in the history of British sports car manufacture rolling off the line, many

125

believed it was only a matter of time before his dream would be realised. In truth, the original outline MGB designs were for a fixed head from which the roadster evolved! Thornley described the MGB GT to the *Motor* at its launch in these words: 'We've produced a motor car now in which no managing director would be ashamed to turn up at the office.'

Although some 220lbs heavier than the soft-top, the GT's improved aerodynamics gave an increased top speed with an unexpected bonus of even better handling, the extra weight of the roof and tailgate further counter-balancing the engine. Demand for the GT was so great that an additional production line had to be set up at Abingdon, making a total of seven. Output from Abingdon would probably have hit an all-time high but for one factor— demand in the United States for Austin-Healeys had slumped still further!

The second major modification to the MGB came in 1967 with the introduction of the long awaited 'fully synchromesh' gearbox, with optional overdrive. Although the latter had been available since 1965, improving both the safety of overtaking manoeuvres and cruising economy, it did nothing to assist the characteristics of the transmission in heavy traffic. The gearbox, being derived from the MGA, had required the vehicle to be stationary before first gear could be selected, and led to general criticism of the car in heavy traffic. Due to the nature of this advanced gearbox, and a desire to provide an automatic transmission option, the monocoque shell was given a wider transmission tunnel and 'MkII' designation, which coincided with the launch of a more powerful alternative model.

The MGC
The introduction of the MGC was the solution taken by BMC to a number

Paddy Hopkirk at the wheel of the MGB entry in the 1965 24-hour Le Mans race. The car finished second in its class, eleventh overall. (National Motor Museum)

of factors prevailing during the mid-1960s. The first was that in spite of considerable investment into the Austin-Healey export programme, sales into the United States continued to fall as calls for a mighty MG grew. Secondly a new 'cloud' was gathering: exhaust emission controls. Stringent new Federal laws pointed to the day when thirsty and inefficient power units such as those used in the A-H 3000 would be refused entry into its main export market. Thirdly, data already gathered by the R & D team had showed that the bodyshell of the MGB was capable of a performance far greater than could be presented by even a highly-tuned four-cylinder engine. The last consideration was that, even though the Austin-Healey Sprite had shared the same bodyshell as the MG Midget since 1962, neither had seemingly suffered any apparent market resistance, sales for both exceeding expectations. That BMC's management took the 'rationalisation' solution and sanctioned a new six-cylinder version of the monocoque in both MG and Austin-Healey guise was obvious, and greeted with little astonishment. What *was* surprising was the time it took—many of the original drawings were covered with six-year-old dust!

Whilst the wisdom of producing the new 'hairy-chested' MGB was without question, the project produced one of the most controversial sports cars to leave the Abingdon factory. When Donald Healey saw the plans for the new engine, he washed his hands of the whole project, preferring to see the Austin-Healey name dropped altogether than see the car given a 3000 MkIV production designation! His main criticism lay with the dimensions and weight of the engine. The depth of the block required the front suspension to be radically redesigned, whilst the length caused the bonnet to require a 'gusset'. Built as a cast-iron, straight-six engine, the weight caused the loss of the nimble handling characteristics of the 1800 cc model and gave only a marginal improvement to acceleration and performance. After two unsuccessful years in terms of sales and press, less than 9,000 cars were produced in both roadster and GT body-styles up until 1969. During the short production life of the car, relatively little was done by the factory to improve its performance, although two light-weight prototypes were raced with limited success at Sebring and Le Mans.

No study of the MGC would be complete without mention of either the 'Downton-tuned' car or 'University Motors Special'. Several private tuning specialists successfully modified the MGC, notably a small company operating near Salisbury. Downton Engineering soon built up a reputation with the car which remains, some 20 years on, legendary. The small workshop was so busy that the company took additional premises in London, where customers were plentiful. One of their clients was University Motors, the main London distributor. In an attempt to clear the lines when production of the MGC ceased, the factory sold the remaining cars to U.M. as a job lot.

One of the two ex-Works 'Sebring' MGCs prepared for the 1968/9 season. This MGC quickly gained a cult status and has spawned dozens of replicas over the past 25 years. (The RoadRunner Collection)

It was their intention to raise the car's specification to an almost bespoke level. Offering a bewildering choice of metallic paintwork and structural modifications, the garage also planned to use extensively the tuning services of Downton Engineering to produce a highly personalised supercar that would sell at a premium. The project failed to attract sufficient interest and the majority of this final batch were sold off, not as 'UMS' cars, but in standard form . . . at a discount!

Ironically, it is the MGC in general that now commands the highest value of any MGs of the period, especially in tourer form. MGCs having physical evidence connecting them with Downton Engineering are quite sought-after, whilst documented 'University Motors Specials' are even more so. Logically, Downton-tuned 'UMS's are by far the most valuable of all!

'Middle-aged Spread'

The merger of British Motor Holdings and Leyland Motors late in 1969 resulted in MG becoming part of the Austin-Morris division of British Leyland Motor Corporation. As such, it lost its own drawing office once again, the Longbridge designers providing a major cost-saving 'face-lift' programme rather than one of reinvestment for the remaining Midget and MGB models.

Giving the 'new' MGB vinyl seats instead of leather, imitation alloy 'Rostyle' wheels and fashionable alloy sports steering wheel, the 'MkIII' was given a dramatically revised grille. Breaking over half a century of tradition, the chromed radiator grille was replaced by a moulded black plastic recess and tiny octagonal badge. Whilst it gave BL a primitive form of corporate identity with other models in their range, and undoubtedly a cost saving, it was a marketing disaster. Such was the outcry that a more acceptable 'honey-comb' grille was fitted. Although still mainly of black plastic and stainless steel, it did utilise the design of the former frame and badge. With minor modifications, the 1800 cc MGB remained in this form until the introduction of the next production phase, the polyurethane or 'rubber-bumpered' model, but not before an exciting stablemate had been launched.

MGB GTV8

When the MGC failed to fulfil the role as successor to the Austin-Healey 3000, a temporary halt was put on the search for a replacement 'hairy-chested' sports car. Private enterprise once again intervened, this time when an auto-engineer, Ken Costello, began fitting the Rover V8 3.5 litre lump into a standard MGB bodyshell. Several dozen 'Costello' hybrids had been built up before one was bought by British Leyland, the supplier of the engine, for 'appraisal'. Supply to Mr Costello dried up as development work on MG's own version began. A prototype existed within a month and a production model was built by December 1972. Eight others quickly followed in LHD form for testing in the United States. It was, however, denied the lucrative American market—not by the nation's regulatory body but by pressure from BL's concessionaires of Jaguar 'E' Types and Triumph Stag models, frightened for their sales! Released onto the English market in August 1973, the MGB GTV8 was the fastest production car in the world, with a top speed in standard form of 135 mph.

The post-War cry of 'export or die' had created millions of jobs in a motor

Two 'Limited Edition' MGB roadsters. On the right, one of the 420 English models finished in bronze with a gold stripe; and, left, the black American version with silver stripe plus alloy wheels and a boot rack. (The RoadRunner Collection)

industry conscious that the major market for its product lay across the Atlantic. Initially, 'feedback' was restricted to style and fashion, but later began to suggest areas of environmental responsibility with exhaust emission controls, via the National Highway and Traffic Safety Administration. Passenger safety was added to the list and quickly developed into an obsessive demand for what Roy Brocklehurst later described as 'a sort of padded cell'. Following a directive insisting on certain criteria being met by 1975, the MGL underwent what was, until 1992, its final facelift.

The 'Rubber-bumper' MGB

Of all the safety features required by the American NHTSA, the ability of a car to absorb an impact of 5 mph without damage was the one visually to affect the MGB most of all. Only with the addition of reinforced beams concealed in heavy polyurethane bumpers at both front and rear (which add 70lb in weight and 5″ in length), could this be achieved. To the dyed-in-the-wool enthusiast, the result was both disfiguring and ugly. When other enforcements such as increased ride height and tightened emission controls were added to the equation, acceleration, top speed and handling all suffered. In spite of all this, the MGB, now in its thirteenth year of production, took on a new lease of life and continued to outsell its would-be successor, the TR7. By the summer of 1979, and in line with most other exporters, British Leyland struggled to retain exports as sterling continued to strengthen against the US dollar. As lines of unsold MGBs stood like forlorn soldiers in a losing battle, BL held out the white flag and surrendered. The last two MGBs to be built at Abingdon were driven off the line on 22 October 1980, destined for the British Motor Heritage Museum.

Production Specials

Although during the long production run of the MGB several prototypes were built, there have been only two 'special' models, sanctioned to commemorate specific occasions at Abingdon; the 'Jubilee' and 'Limited Edition' models.

The 'Jubilee' was available only in MGB GT form during 1975, a gesture in celebration of 50 years of MG production. All were finished in British racing green with gold V8-style wheels and stripes along the wings. Only 750 models were commissioned, a number accepted as representing one for each BL main dealer in the United Kingdom. Each first registered owner had the option of having his/her name inscribed on a brass dashboard plaque and was given a commemorative watch.

Exactly 1000 'Limited Edition' MGBs were built to mark the end of MG production in 1980. A total of 420 roadsters were finished in bronze coachwork, the remainder being GTs finished in pewter. Both were available with either wire wheels fitted or special bolt-on alloys and all given a distinc-

The rare sight in 1992 of four 'Jubilee' models complete with gold wing transfers and matching V8-style wheels. (The RoadRunner Collection)

tive front spoiler and side livery.

There also exists an unknown number of US-specification 'Limited Edition' roadsters . . . a statement that many would argue is a contradiction! In an attempt to ease the plight of the American MGB importers, MG produced and sent out hundreds of 'LE' packs for use on LHD MGB roadsters, which included silver side-stripes, a boot-rack, alloy wheels and other minor cosmetics. Unlike the home-market 'Jubilee' or 'Limited Edition' models which were completed at Abingdon, it was the responsibility of the agent to fit the kits to the cars already in stock. This inevitably means that the actual number of 'modified' cars is unknown and the standard of workmanship variable. As the vast majority of exported MGBs were finished in black, it would be logical to suggest most US spec LE models would be found in that colour.

Rebirth
When the Leyland management took the decision in 1979 to close the Abingdon factory and production of MGs ceased—despite the storm of protest in Britain and from the rest of the world—there were those who believed the marque would die and active support fade away. In reality, the opposite happened and values of the sports car, now out of production, climbed. New businesses sprang up, remanufacturing unavailable parts, spurred on

131

Brooklands in 1991 is the setting for this MGA rebuild entered in the retrospective Monte Carlo Rally. The restoration was authentic, right down to a period hardtop and luggage on the bootlid! (The RoadRunner Collection)

by enthusiasts wishing to keep their cars not simply roadworthy but in as-new condition. Responding to the call, the custodians of the old BMC/Leyland marques (then grouped as Austin-Rover) formed a remanufacturing and market arm of their organisation. This body, named 'British Heritage', embarked on a bold business strategy aimed at turning back the clock.

Following a reconnaissance of almost every storage area of every plant belonging to Austin-Rover, over 95 per cent of the original MGB tooling necessary for production of a complete bodyshell was recovered, much having stood outside for almost a decade. Once cleaned and re-furbished, it was adapted and installed into a small industrial unit. Using a hand-picked team of skilled technicians, some ex-MG production of the MGB roadster body-shell began again in 1988. Over the next four years, literally hundreds of MGBs previously considered unrepairable and destined for the scrap yard, have been saved and rebuilt using a 'Heritage' shell, many by the DIY enthusiast in his garage.

(Above) The latest MG, the RV8, with its famous predecessor, a 1962 MGB.
(Below) The cockpit of the RV8 is undoubtedly the most luxurious ever
seen in an MG!

Production of the MG RV8

The greatest sign of the renaissance of the marque has, however, been the launching of a new MG, the RV8. Code-named 'Adder', the RV8 is a rich blend of traditional engineering experience and modern technology, marketed by Rover Special Products. Using a monocoque front the 'rubber-bumper' era of the roadster and many of the original car's basic principles, the development team at Rover's Canley home have created a supercar, both in terms of performance and passenger comfort. Body-styling is dynamically different yet reassuringly familiar; sculptured headlamps and tail-light clusters understating the case. The handling, although improved by a wider track, lower stance and the latest telescopic shock absorbers all round, still relies heavily on 'cart' springs for support, rack and pinion steering rack and vacuum servo-assisted brakes. Using a power unit similar in design to the 1973/6 model, the 1992 V8 configuration is fuel-injected and fitted with catalytical converter and mated to the latest five-speed gearbox, eliminating the need for an overdrive facility.

When the RV8 was launched at the Motor Show in October 1992, it attracted mature members of the public who had never been to a Motor Show before, drew crowds fifteen deep around the stand and so much enthusiasm from the motoring fraternity that even Rover executives were visibly shocked, being under siege for the duration of the show! Underlining the statement made in the opening paragraph of this chapter, the MG RV8 is the latest demonstration of the MGB's undeniable ability to move with the times, adapting to the changing trends of the buying public. Whilst all applaud Rover's decision to bring back a revamped MGB, a considerable number question its success with a £27,000 price tag and 3.9 litre performance on our rather restricted road network.

In truth, the decision by the Rover Group to launch in 1992 a version of a car designed over 35 years previously is unprecedented in the history of the industry—possibly even the wheel! That is not to say that the MGB is the greatest car of all time—rather one whose popularity refuses to go away.

Restoring MG Cars

Mike Allison

Once bitten by the MG bug, the ultimate ambition of all enthusiasts soon becomes to restore an example of their own favourite model—if they can find a suitable example, that is. Indeed, every contributor to this book has his, or her, favourite story to tell of a restoration triumphantly completed—or one sadly aborted and turned over to the professionals. Among MG restoration specialists, Mike Allison is a man held in high regard—and for good reasons. Mike worked for MG from 1962 to 1973, latterly as Chief Quality Engineer. He then opened a garage, restoring MGBs and Midgets, broadening his scope five years ago to include prewar MGs, too. To this wealth of practical experience, he can add the claim of having driven every type of MG except the R-type and the 6R4! In the following pages, Mike offers a wealth of sound advice to anyone contemplating restoring their dream car. He is as direct about the problems as the pleasures, and gives advice that will prove of value to both the expert and the layman (or woman!).

So you are going to buy and restore an MG? Good! It's a super hobby which I have enjoyed for more years than I care to admit, and one which has brought my wife and me a worldwide circle of friends. You will be a welcome addition to our ranks and we look forward to meeting you with your car in due course.

MG cars have always been an attractive proposition for the amateur restorer because the method of construction was simple, which means that the cars are easy to take to pieces and to put back together again. It takes very little to raise enthusiasm for the ownership of an MG and if enthusiasm is given its head it is all too easy to think of a car as a restoration project to occupy one in the winter or in one's retirement as an attractive way of spending time, resulting in something which can be enjoyed for a long time to come.

So let us start off by dampening enthusiasm: more classic car restorations have been started than have been finished, and there have unfortunately been a large number of MGs amongst these. We will therefore begin with a set of advisory notes to ensure that your car does not join the ranks of unfinished projects, and with you as a disaffected enthusiast.

First and foremost, do not think that this project can be done cheaply. It will absorb money at a frightening rate, especially when you seek the help of the professionals. Professional help is essential at times, and please remember that these people stay in business to earn a living. They are not interested in your personal financial problems; they have enough of their own, looking after taxes, rates, telephone and power bills, not to mention rents, materials and sub-contracted services such as pay for employees! The actual cost of the job can be kept low the more work you do for yourself, but, even so, at the end of the job you will have spent quite a lot on parts and services. However, this is your chosen hobby, so you should not expect to have this paid for by others, and whatever you do will cost something. Remember to pace your spending at a rate which you can afford—and don't let enthusiasm overtake your pocket!

One last point before we start to talk objectively: do not ever think that a restoration project will earn you money. It positively will not. You may be lucky and break even in financial terms, but the cold fact is that, when you have finished, it will have cost a lot more than the value of the car at conclusion of the restoration *in real terms*, taking labour costs in account. If it has not, then you have skimped the job!

Selecting the Car
Buying a car is fairly easy, although you will have to start by making sure that you do not make the serious mistake of buying a car you cannot bear to drive! Cars are meant to be driven, but if you buy a car because it looks nice and then cannot bear to drive it, this is a great shame. I actually like driving MGs and have driven examples of most models, and can say that some satisfy me while others do not. You may differ in your own opinions, and it is all very well hearing me extol the virtues of, say, a J2, but if you find that a hard ride and ultra-light steering, coupled with a braking performance which is much lower than current standards, is unacceptable to you, you are going to hate every minute in the car . . . however nice it looks in your drive.

The best starting point is to join one of the clubs, and befriend various owners at club meetings and try to find why they like their cars. Even beg rides in various cars to see if you can get some ideas of what it is all about. Do not rely on the opinions of one owner or one car, either. I like the J2, but I've driven a couple of lousy ones! If my enthusiasm had been based on these

Rebuilding an MG is all about the end product—but never forget what goes between the idea and the finished article shown here.

I might still have been involved in stamp collecting! If possible, drive a number of different models which attract you and finally make a decision as to which model you are going to try to buy. Don't be too fixed in your ideas. You may decide on a J2, but find a TC or something which actually suits you better. Make sure that you have tried a TC (or whatever) before buying!

Next, make sure that your partner in life likes the idea: I know of several MGBs and Midgets bought 'for the wife', but the lady hates driving the thing. This, of course, means that you have to use it as daily transport, which may have been your ulterior motive, but it does bring up another important point. All the so-called classic cars have been superseded, and 'modern' cars are the best vehicles for pure transport in terms of comfort and economy. I love driving my 'N' type Magnette and have driven it to most parts of the UK, but if I have *got to go somewhere* I drive the 'Eurobox' which shares the garage . . .

Then, again, consider the carrying capacity of the car of your fancy. If you have a young family to transport, a J2 is definitely not for you, and you may think that a four-seater J1 fits the bill better. Again, do try one fully-laden first. One afternoon in the back seat of an open car and the family may well make you decide to buy a Z Magnette!

Necessary Skills

If you have never rebuilt a car, do not start off with something which is a mechanical challenge unless you are an engineer—or are prepared to spend a long time learning. I would advise against starting with an early car unless you have a real enthusiasm for this. Start with a post-War car from the Fifties or Sixties which was made in larger numbers, and for which spares are easy and economic to buy. Fortunately, spares for MGs *are* fairly easy to buy and are reasonably priced, but these terms are relative! At the time of writing, you are unlikely to rebuild a car for less than £5000 in bought-out services and new parts, whatever model you choose and however much you do yourself. The older the car, or the more jobs you farm out, the higher this price will be.

If you are not good with your hands, do not even bother to start. You will have to learn a wide range of skills, some of which take a lifetime to learn and at which you will only get proficient with constant practice, paint-spraying and welding being two such skills . . . Having said that, provided you are not in a hurry, you will be able to make a passable job if you really do learn as you go along.

To be a good car restorer you will need to be a good general mechanic and a competent fitter, which are NOT the same thing. In addition you will need to be an electrician, a sheet metal worker, a turner and machinist, a welder, an upholsterer, a paint sprayer and, of course, an expert on the car of your choice. If you are going to restore a vintage model, in addition to those skills already mentioned, it will be necessary to have an appreciation of coachbuilding skills and be a good carpenter. You will also have to have a deep appreciation of basic engineering and be capable of sorting out a problem from first principles, which really means understanding mechanics and mathematics. The older cars require a great deal of patience when assembling them, and parts often need careful hand-fitting to make them work, especially when they are made to a drawing!

Lastly, do not attempt any job which you do not think that you can complete. If you do not think you will be able to do something, the chances are that you will not. If you start to work on a component and then fail, you will not be greeted with much sympathy if you arrive at a professional workshop with a load of broken bits which you have damaged during unsympathetic stripping down!

If I have not yet persuaded you that restoring an MG is not for you, then we had better get down to the job in hand. The first thing you will need is somewhere to work, and some tools to do the work with.

Always lay the parts out for a sub-assembly before starting, to make sure that you have all you need for the day's work.

Essential Workshop and Tools

While a garage in which to work is not strictly necessary, it does make life more bearable, especially in the winter. I have never done a restoration without having a garage, but when I was young I did perform engine overhauls in dad's garden shed, having removed the unit on the driveway outside. Your garage will ideally have enough room to park the car in and to be able to walk right round it, preferably with the car doors open. A space of around 16 feet square is good, and the larger the working space the better, but you will be ruled by what is available at your home. Just as you do not *need* a mansion to live in, it is possible to carry out a restoration in the garage attached to a normal house.

A strong bench is essential, about three feet high. A surface two feet deep and five feet wide is as large as you will need, with a four- or five-inch mechanic's vice mounted at the right-hand end (assuming you are right-handed) and mounted fairly close to one of the bench supports. Unless you have around ten feet of bench length, do not mount the vice in the middle! If you are making your bench, the ideal height is for the top of the vice jaws to be at elbow level when you stand beside it. The bench should be very strong, preferably cast-iron framed, and with legs at not more than four-foot intervals. The top should be of wood at least one inch, preferably two, thick and covered with aluminium sheet to protect the wood from damage. The bench should be at one end of the garage and not along one of the side walls. Provide yourself with racking and storage for parts and part-finished work, and use a felt-tip marker for identification of these and the state of restoration. Good lighting is essential: end-to-end continuous strips of fluorescent lighting at five-foot intervals across the roof are ideal, and in addition a portable fluorescent tube is useful to light those jobs which are obscured.

A really good set of spanners, open-ended and ring, is absolutely essential and a socket set is highly desirable. These should be either Whitworth or American convention, according to the car you are rebuilding. You will need a number of screwdrivers, pliers and a couple of hammers of different weights. But other tools can be bought as you need them. All I will say is that you should buy good quality tools which will probably last you a lifetime. They are expensive, but worth every penny. I still have and use tools which I bought or had as presents when I was a young man, and many of these are used every day.

Do not make the mistake of buying expensive capital equipment, such as paint spraying gear, and lifting tackle: all this can be hired for a reasonable fee. If you buy it, you will find it blocking up a limited working space for most of its life, and it will never earn its keep. A rotary grinder, equipped with a wire brush at one end, and a good electric drill together with a set of high-speed steel twist drills are pretty invaluable, but few other machines are truly essential and can easily be bought as needs dictate. A lathe and milling machine can save pounds if you really know how to use them, but otherwise they are expensive luxuries.

Starting Operations

Do not, whatever you do, take the car to pieces and reduce it to a mass of small parts. This is the way of the true amateur and is almost certain to result in you deciding that the mountain of work can never be finished!

It is better to take stock of what work is needed. Take as many photographs as you can in good light or with the benefit of flash, which not only show the general layout, but also look critically at the detailed work of the car. This is most important, since you will almost certainly forget what goes where and what the order of assembly of parts is. I usually try to take notes with sketches. Compare what you have with photographs of factory origin, but beware of being too critical of non-standard parts. If you can determine that non-standard parts are period, it may well be worth retaining them. Too many cars have been returned to factory standard, while quite a few were modified when new by owners or even by the factory at the owner's request. Some were updated by dealers when new, to sell 'last year's model'. The non-standard parts may therefore be part of the history of the car. Car club experts can give you guidance on these matters, and so can other owners.

There are two basic methods of car construction which will predetermine your procedure during the rebuild. All MGs built up to 1955, except the 'Z' Magnette, were built on a chassis frame, as of course was the MGA, which finished being built in 1962. All other MGs are built on the monocoque principle, in which the chassis is integral with the bodyshell. I will separate the differing methods inasmuch as they influence the rebuild procedure, but the basic method of rebuild does not actually differ very much.

Cars with Chassis

To rebuild one of these cars, the best way to start, having taken your photographs at the start of the project, is to dismantle into major components and then attack each of these in sequence as suggested below.

Personally, I think it is logical to start with the chassis frame, and then to progress to the suspension, the axles and suspension, next the steering and brake systems, then to work on the body and trim, finally rebuilding the engine and gearbox.

CHASSIS AND SUSPENSION

The chassis frame has to be scraped clean of oil and grease. A solvent cleaner, such as 'Jizer' or 'Gunk', is an essential aid to this job. I have used a garden spray for this purpose, which removes much of the drudgery from this dis-

It is logical to start with the chassis frame.

tasteful operation. Once all the gunge is removed, the frame can be checked to make sure it is 'square'. To do this, place the frame on a level surface, and measure across the diagonals of the frame. These should all be equal within 1 per cent of each overall measurement. Frame members should not be visibly bent. Surface corrosion is acceptable, but if the corrosion has reduced the local thickness of the frame material to less than 60 per cent of its original, then that part of the frame should be replaced by welding in pieces of steel of the original thickness, and smoothing off to a good finish. I personally do not approve of new frames, on the basis that they do not form a true restoration job, but if your frame is very badly corroded this may be the more economic course of action. To build a car from scratch on a new frame is definitely 'cheating', however.

Once the frame is pronounced true, it can be sent away for shot blasting and painting. But before this is done, the threads should be protected by inserting plugs into them. I prefer plastic powder coating as a treatment for the frame, which is permanent, but the original frames were treated to two or three coats of chassis black, which is a self-priming oil-proof paint. All chassis components should be treated with this paint.

Suspension components, such as springs and their pivot bushes or trunnions, should be inspected, but in general are replaced with new parts. I am not in favour of binding springs with cord or gaiters. It looks very nice when newly done, but this does tend to hide problems when the car is in use and, more importantly, it stops the springs doing their job! The original purpose

Having mounted the front axle, fit the brake gear—the first safety point of 'Safety Fast'.

of binding was to do just this when racing at Brooklands, but has no place in the modern world.

AXLES AND BRAKING SYSTEM

Surprisingly, some of the least likely components may be in need of careful and skilful attention. MG front axles are often bent or twisted and should be checked and re-set before they are prepared for refitting to the chassis. Rear axle cases are also often bent, so they should also be carefully checked. The threaded ends may require restoration, which, like the other work in this paragraph, will require professional assistance.

Kingpins are simple to replace, but need the correct sized reamer. In my opinion, they are far better honed on a machine, so again your friendly specialist will need a visit. Please do not hammer the bushes into the swivel pin housing. Have them pressed in: with a little ingenuity you can use your bench vice as a press.

Brake gear comes next and as there is no more important part of a car, special care is needed. I personally do not favour skimming brake drums; MG drums are far too thin for this. Mild scoring is of no consequence, but if the drum is grossly out of round, it will need either replacement or building-up with metal spray and restoring to its original diameter. If you decide to replace a drum, replace these in axle pairs. The shoes will, of course, be relined. For road use, I favour a soft lining as originally specified. It may be necessary to file linings to get them to be a nice fit in the drum, and to give a good contact with the drum as the brake is applied. In any case a bevel or 'lead' should be filed on the leading edge of each shoe.

Hydraulic brakes are straightforward in that most of the necessary parts are available. As a matter of course, the hoses should be replaced, and I would replace the master and wheel cylinders, too. If the main pipes are to be replaced, it is better to use 'Kunifer' material, rather than steel bundy, and copper is taboo because it hardens in use.

If you are dealing with mechanical brakes, then you will need to replace pivot bushes, and the operating cams will need some work. Cables will need replacing with new ones. If there are any rod connections in the system, these will need a thorough overhaul as determined by their construction. When the system is all reassembled, aim for zero friction. It is impossible, of course, but everything should operate smoothly with little effort. With light hand pressure on the brake pedal, it should be possible to stop a wheel rotating quickly.

Having reached this stage, it is time to fit in the electrical harness or loom, making sure that it is clear of any possible moving mechanical parts which will chafe through and cause a fire. I like to use neoprene rubber-insulated 'P' clips for this job.

BODYWORK

The bodywork is the next major job, and inevitably you will need a wooden frame. If you are a good carpenter, then the chances are that you will make a reasonable job of the body-frame, but I am one who will go for the ready-made frame every time. These are available for all the two-seater MG models. Those with four-seaters or saloons are on their own here, although you will always find a professional coachbuilder who will make you a frame. The correct wood is seasoned ash, and almost nothing else is suitable; most other woods will split or break up when the car is used. Floorboards are best made in marine plyboard of 9mm or 12mm thickness. All the woodwork can be treated with wood or preservative, but I prefer painting. The metal panelling is fairly easy to do, but you will certainly need the help of a professional for the bonnet, wings and scuttle top.

The wings and bonnet can generally be restored for less than it will cost to buy new and fit them. Originally the wings were individually fitted to each car, and it takes quite a lot of time to do this so that the car looks complete

When you get to the body stage, make up templates of its eventual shape to ensure the bulkhead and firewalls are in the correct line, and that the body is level.

145

and finished. It is important not to skimp this job, as these are the most prominent parts of an MG.

Once the bodywork is complete, it is sensible to get it painted, normally before getting it trimmed. It is possible to do this yourself, but is an anti-social occupation which may not be viewed with great favour in some clean air districts. Check before you start. If you are doing the job yourself, hire a good quality spray kit, and use cellulose paint. If you are having the car professionally painted, you will be told that 'two-pack' or epoxy finish is more durable. What this really means is that the paint shop is equipped for this process. Whatever process you use, your car will inevitably need repair work carried out from time to time, and the fact is that cellulose is much easier to repair at home than two-pack. Against this, cellulose finish will chip more readily, so you make your own decision here!

Before the car can be trimmed, the instruments will require fitting, and the electrical components fitted and the wiring completed. Then all sorts of detail work will need completing at this stage. Do not be tempted to put minor jobs off. Cars advertised for sale as 'Major work completed, just requires finishing' hide a reality that there is a wealth of detail work which will actually cost almost as much to finish as a complete rebuild—especially if parts are to be found or made up. All jobs should be attacked as they are needed, otherwise the cost of doing them later can become an enormous burden.

Trimming is a fairly easy job on two-seater MGs because it is possible to buy a complete trim kit, which is fairly easy to fit, except for the hood and tonneau. Saloons and four-seaters will require the services of a professional. Do not even think of doing the job yourself unless you have an industrial sewing machine, and know how to handle the expensive material. It is all too easy to mess them up!

ENGINE AND TRANSMISSION

While the car is being painted and trimmed is the time to think about the engine and gearbox. I leave the engine until fairly near the end, chiefly because one is not then tempted to start it before being able to drive the car. Repeated starting and stopping a cold engine is one way to ensure that the cylinder bores and pistons are worn out in a short time!

If the engine originally had white metal bearings, these will need replacing. Please do not be tempted to change them to shells; this is a blind alley unless you have a fully equipped machine shop and the skills to use it! The temptation to use shell bearings is based on the assumption that they are cheaper to replace than cast and machined bearings. This is true, but in my experience bearings last around 10 to 12 years, covering quite a high mileage, and about the same as the life of the pistons. The cost involved in converting to shells

is such that it would not be recouped in 70 years of ownership of the car!

Spend the money on making sure that the engine will last. Have the crankshaft and connecting rods crack-tested. If the news is bad, you will need to buy a new crank and rods set. This will be a good investment in the long run, but is expensive. Much more expensive, however, is taking a chance and using a cracked crank or connecting rod!

Always rebore an engine which is undergoing a complete overhaul. Do not be tempted to use old pistons, which is a false economy. The assembly of an engine is fairly simple, but do make sure that everything is clean—and clean it again a final time before assembly. Dirt is the enemy of an engine, so do not skimp this operation. Oilways should be cleaned out, especially those in the crankshaft if you are re-using an old one. If you ask the regrinding man he will remove the plugs for you, and he may even be prepared to make up new ones to fit after cleaning, which you should fit.

CYLINDER HEAD

Work on the cylinder head in the case of the overhead camshaft models is a

147

A PB crankcase assembly ready for fitting the cylinder head: all this work
is much easier to do out of the car.

lengthy business. But whichever type of engine you have, careful work at this
stage will determine the way your engine runs in the future. It is important
that valve seats are in good condition, not cracked, and that they are not
recessed into the head casting. Various repair methods are available, but I
would advise against welding to repair, which can be an expensive way of
reducing your head to scrap. The cracks can be repaired using a cold 'metal
stitch' process, and inserts restore the seats themselves. I have successfully
used this method of repair for years.

Fitting valves is straightforward. Always use new valves and springs, and
lap them in to give a seat 1/16" wide. Cam followers or rockers can be
stoned to give a smooth contour, but gross wear will have to be overcome by
replacement. The cam follower to camshaft relationship is the secret of
making an OHC engine 'go', and money saved here will only result in poor
performance. With the pushrod engines, the rockers can be rebuilt by hard
surfacing, but the base tappets should be replaced. Camshafts should be

replaced if worn. Much of the work in reassembling a cylinder head is labour intensive, but there is considerable skill involved with the OHC engines. Most of the knowledge is well-known in the appropriate Register circles, and information on 'how to do it' has been written up many times.

BALANCING

Always have the crankshaft, complete with flywheel and clutch, connecting rods and pistons balanced, which will help the longevity of the engine. In addition, I would advise having a competent mechanic balance the combustion chambers and polish the ports of the head. A word of caution on this second item; beware the man who talks glibly about 'gas-flowing'. What you really want is for the combustion chambers to be equalised using a burette, and if our man knows about this, the chances are he will also know about optimum shapes. A good man will insist on fitting your new valve guides, and valves, at this stage.

Fitting pistons using a ring compressor avoids breaking the rings. This one is made from a strip of steel rolled into an overlapping cylinder which is compressed using an ordinary Jubilee clip.

Gearbox

The gearbox is often treated as an item of mysticism. In fact it is a simple piece of equipment, but does require skills which you may not care to acquire, especially in the case of an automatic or preselector. In this case you will need the help of your specialist friend; be wary of the man who advertises in the yellow pages to overhaul 'all types' of transmission. It is better to deal with someone who actually specialises in working on old car transmissions!

Monocoque Construction

These cars include all models from the 'Z' Magnette on, except the MGA. The major restoration work will be in the body unit, which will include a lot of sheet metal work and welding, followed by paint spraying and rust proofing, and then trimming. This work is actually the major part of restoring this type of car.

Bodyshell

It seems tempting to buy a new bodyshell where these are available, but, in my experience, unless you are restoring a car of the actual model year for which the new shell was built, you will find it more cost-effective to restore your old bodyshell. Each model year of production of the MGB and the Midget resulted in revised body details, which means that parts are not necessarily interchangeable from year to year, while the new shells are made to suit a sweep of model years. All this adds up to additional hassle which the amateur car restorer can do without, especially if he is genuinely interested in *restoring* his car. If you want to convert your rusting hulk into something usable, then that is a different story.

Professional restoration of your old shell should not cost more than a new shell, and you have the benefit of knowing that you end up with the 'original' car. Of course you may have decided that you can do this yourself, and bought panels to weld in place. Make sure that you fully understand the construction of the body before undertaking this sort of work, however. It is definitely not for the faint-hearted!

Mechanical components

After work on the bodyshell, and once it is painted, you will be able to start on the mechanical work. If you have restored the units separately as time was passing, you will be able to fit them onto the bodyshell in a matter of about 40 hours' work.

The mechanical components are tackled in sequence in much the same way as was described for the older models, so there is no point in repeating the details. The really good thing about working on one of these cars is the proliferation of dealers who can supply parts to you, and there will be no

An N-type engine assembly nearly ready to start for the first time—the most exciting moment of any rebuild!

long delays in waiting for parts at all. Some of the major units may well be available as overhauled exchange items at a fraction of their rebuild cost. Provided you have the money available, it is quite feasible to rebuild an MGB or Midget over a winter, once the body is complete and ready to accept the mechanical parts.

Restoring one of these later model MG cars really is a satisfying project which it is possible to complete within a short time frame, and the finished project should be considerably cheaper in financial terms than the work on an early car.

Tuning for Extra Performance
One of the things I am most frequently asked about is tuning an engine, by which the owner usually means 'super-tuning'—that is, making the car go

faster than it did originally! This is fairly easy for an MG and is helped by the wealth of information which the factory released at the time of construction. However, one of the problems one has to face with an old car is that it will not go as fast as the modern 'Eurobox', even when the former is in a comparatively high state of tune. This, I am afraid, is a fact of life: performance standards improve with the passage of time. A K3 Magnette, a specialised racing car, was a very fast car by the standards of 1933, but by 1963 the MGB, a mass-produced road car, was nearly as fast and much more reliable and comfortable to ride in. Now, a GTI saloon will see both these cars off comprehensively: but what is this proving? If you want an old car it must be for the intrinsic satisfaction which ownership gives you. Certainly the performance of the car must be satisfying, but this must be relative to the time of build of the car—not today!

In order to increase the power output by, say, 20 per cent, the reliability of the engine may well suffer to a level where driving the car will be a chore, and no fun. In general, therefore, I would advise against tuning the car to go faster unless you want to go in for competitions, which is another story altogether. The fact is that a standard MG will cover ground fairly quickly and it will be possible to average 30/40 mph whichever model MG you have. To average higher speeds one should really be thinking in terms of a much faster car equipped with radar detection equipment!

The next most frequent question I am asked is about raising the rear axle ratio, normally to obtain more relaxed cruising on motorways. My immediate response is: why drive on motorways? All the pre-1955 MGs were built before motorways were thought of, and were given a final drive ratio which would provide a good road performance. If you raise the ratio, then the acceleration will suffer, possibly to the stage where the car is painfully slow. Don't forget that MGs are fairly heavy cars and an engine output increase of 20 per cent may be necessary to overcome an increase in final drive ratio of two teeth, which is the minimum! Now re-read this section, and you will appreciate that we are in a Catch-22 situation.

If you stay off the motorways, the journeys are much more pleasant—and the roads more fun to drive. My own feeling is that a standard MG is nice to drive. Why change it?

Finishing the Project

Once the car is finished, comes the most exciting moment: starting it up and driving it. But before even thinking about this, get the car insured at least for third party risks. This will not be a cheap operation, but is an investment which is worth having. Ideally, you will never need the services of the insurance company, but in my experience it is well worth having a policy from a first class broker who will give you the advice you require to get the best cover for your needs.

Starting up is a time for a party, but do try it in private first! Almost certainly there will be a couple of water and oil leaks to cure. If the car has hydraulic brakes, these will need a few minor adjustments to make them function properly. If you have cable brakes, there will be quite a bit of fun making them work properly, and pulling the car up four-square! If you have a car-starting party, make sure that the car part is over before the party. Cars run very well on alcohol fuel, but drivers don't.

Next, you will have to comply with the law, and have the car tested. Your friendly MOT station will normally adjust your headlights for you, and possibly adjust the tracking for a small additional charge, but everything else must be in full working order, so make sure that your brakes work, and all the electrics function properly. Indicators are not necessary on cars built before 1953, but if they are fitted they must work: non-functioning trafficators are not allowed! If you have fitted flashing indicators to an early car which work in the side-lamps and rear lights, in addition to the flashing light, they must also show a constant light from the same point, so double filament bulbs are necessary. A brake lamp at the rear is permissible in the UK as the indicator, but this will involve the use of a flash inhibitor circuit, and you will need the help here of a really good auto-electrician.

Once the car is tested, you can tax it and use it legally on the highway, after which joy should be unbounded! *Then* is the time to think about the *next* project!

Every MG enthusiast dreams of owning a genuine works racing car. This one is a Q-type, one of just eight built in 1934. Dream on!

The MG Models

1 *MG 11.9 hp 'Raworth' Sports*
Six of these 11.9 hp two-seater sports cars with bodies built by Charles Raworth of Oxford were the first to be made and sold by the embryo MG company between the summer of 1923 and late 1924. They cost £350 each. The four-cylinder, side valve engine with mushroom tappets had a capacity of 1548 cc generating about 25–30 bhp. The gearbox was three-speed non synchromesh, and the wheels bolt-on artillery types with 9-inch drum brakes on the rear only. Bore & Stroke: 69.5 × 102 mm.

2 *MG 14/28 Super Sports*
About 400 two-seater, four-seater and various closed versions of this model were manufactured between the winter of 1924 and late 1926, selling at £460. The four-cylinder, side valve engine with mushroom tappets had a capacity of 1802 cc generating approximately 30 bhp. The gearbox was three-speed non-synchromesh with bolt-on artillery type wheels with Ace discs from 1924–25, and bolt-on wire spoke from 1925–6. The model had 12-inch drum brakes front and rear, servo assisted. Bore & Stroke: 75 × 102 mm.

3 *MG 14/40 Mark IV Sports*
An estimated 900 cars—two-seaters, four-seaters and various covered versions—were built between the autumn of 1926 and late 1929. The 'Mark IV' designation is believed to refer to the fact that the MG was then in its fourth year of production. The Mark IV had a four-cylinder, side valve engine with mushroom tappets and a capacity of 1802 cc generating 35 bhp at 4000 rpm. The gearbox was three-speed non-synchromesh and the bolt-on wire spoke wheels had 12-inch front and rear servo-assisted brakes. Bore & Stroke: 75 × 102 mm.

4 *MG 18/80 Six Mark I*
Between the autumn of 1928 and the middle of 1931, 501 cars including

two-seaters, four-seaters and several covered versions of the Mark I were constructed. It had a six-cylinder chain-driven overhead camshaft engine with a capacity of 2468 cc producing approximately 60 bhp at 3500 rpm. The gearbox was three-speed non-synchromesh with centre-lock wire spoke wheels and 12-inch brake drums. Bore & Stroke: 69 × 110 mm.

5 MG 18/80 Six Mark II

In production from late 1929 to mid-1933, some 228 Marks IIs were built including two-seaters, four-seaters and various covered models including a coupé and four-door de luxe Saloon. The Mark II had a six-cylinder chain-driven overhead camshaft engine with a capacity of 2468 cc generating approximately 60 bhp at 3500 rpm. The gearbox was four-speed non-synchromesh and the centre-lock wire spoke wheels had 14-inch drums; some later models had cable brakes. Bore & Stroke: 69 × 110 mm.

6 MG 18/100 Six Mark III

Just five of these Mark IIIs were built during 1930, all four-seater Sports/Racing models. The cars had six-cylinder chain-driven overhead camshaft engines with a capacity of 2468 cc generating approximately 80 bhp at 4000 rpm. The gearbox was four-speed non-synchromesh and the centre-lock wire spoke wheels had 14-inch cable-operated brakes. Bore & Stroke: 69 × 110 mm.

Spanning the years—'Old Number One' and an 'Immortal T-Type'. (Ron Cover)

155

7 *MG 'M' Type Midget*
The first model to be produced in really large numbers—a total of 3,325 cars, including two-seater sports and Sportsmans' Coupé, were made between the autumn of 1928 and the summer of 1932. The 'M' type had a four-cylinder overhead camshaft engine driven through a vertical dynamo, with a capacity of 847 cc generating, initially, 20 bhp at 4000 rpm and, after 1930, 27 bhp at 4500 rpm. The gearbox was three-speed non-synchromesh and the bolt-on wire spoke wheels had 8-inch cable-operated brakes. Bore & Stroke: 57 × 83 mm.

8 *MG 'C' Type Midget*
Only 44 two-seater Sports/Racing Car versions of this model were constructed from the summer of 1931 to mid-1932. The 'C' Type had a four-cylinder chain-driven overhead camshaft engine driven through a vertical dynamo, with a capacity (in 1931) of 746 cc generating 36 bhp at 6000 rpm; by the following year this had been increased to 746 cc generating 44 bhp at 6000 rpm. The gearbox was four-speed synchromesh and the centre-lock wire spoke wheels had 8-inch cable-operated drums. Bore & Stroke: 57 × 71 mm.

9 *MG 'D' Type Midget*
A total of 250 of these four-seater tourers and Salonettes were built between the autumn of 1931 and the summer of 1932. The vehicles had four-cylinder overhead camshaft engines driven through a vertical dynamo with a capacity of 847 cc generating 27 bhp at 4500 rpm. The gearbox was three-speed non-synchromesh and the centre-lock wire spoke wheels had 8-inch cable-operated brakes. Bore & Stroke: 57 × 83 mm.

10 *MG J1 Midget*
From the summer of 1932 to the early months of 1934, just 380 models of the J1 four-seater tourer and Salonette were built. The car had a four-cylinder overhead camshaft engine driven through a vertical dynamo, with a capacity of 847 cc generating 36 bhp at 5500 rpm. The gearbox was four-speed non-synchromesh and the centre-lock wire spoke wheels had 8-inch cable-operated brakes. Bore & Stroke: 57 × 83 mm.

11 *MG J2 Midget*
The sizeable total of 2,083 of the two-seater sports J2 models were produced between mid-1932 and the early months of 1934. The cars had a four-cylinder overhead camshaft engine driven through a vertical dynamo with a capacity of 847 cc generating 36 bhp at 5500 rpm. The J2 gearbox was four-speed non-synchromesh and the centre-lock wire spoke wheels had 8-inch cable-operated brakes. Bore & Stroke: 57 × 83 mm.

A Product of Enthusiasm!

The 8/33 M.G. Midget Sportsman's Coupe, £245

The **MG** *Sports*

"Faster than most."

The idea of the M.G. Sports car was conceived in a moment of enthusiasm!

It was designed by an enthusiast, is built by enthusiasts and is sold by enthusiasts.

Can you wonder that M.G. owners are themselves enthusiastic?

We cordially invite you to visit the new factory at Pavlova Works, Abingdon, where you can see enthusiasm being built into every car.

8/33 M.G. Midget Sports, from £185   18/80 M.G. Six Sports, from £510

The M.G. Car Company
Pavlova Works
Abingdon-on-Thames

Phone 251 (3 lines)   *Wire: "Emgee"*

1930: An early MG advertisement for the 8/33 Midget Sportsman Coupé.

12 *MG J3 Midget*

Just 22 of the two-seater J3 sports car were produced from the last months of 1932 to the end of the following year. The four-cylinder overhead camshaft engine driven through a vertical dynamo had a capacity of 746 cc generating 45 bhp at 6000 rpm. The gearbox was four-speed non-synchromesh and the centre-lock wire spoke wheels had 8-inch cable-operated brakes. Bore & Stroke: 57 × 71 mm.

13 *MG J4 Midget*

Even fewer of the two-seater Sports/Racing J4s were made from mid-1932 to early 1934: just nine. The four-cylinder overhead camshaft engines driven through a vertical dynamo had a capacity of 746 cc generating 72 bhp at 6000 rpm. The gearbox was four-speed non-synchromesh and the centre-lock wire spoke wheels had 12-inch cable-operated brakes. Bore & Stroke: 57 × 71 mm.

14 *MG F1 Magna*

A total of 114 of these four-seater open and Salonette F1s were produced between the end of 1931 and the winter of 1932. The six-cylinder overhead camshaft engine driven through a vertical dynamo had a capacity of 1271 cc generating 37 bhp at 4100 rpm. The gearbox was four-speed non-synchromesh and the centre-lock wire spoke wheels had 8-inch cable-operated brakes. Bore & Stroke: 57 × 83 mm.

15 *MG F2 Magna*

40 of the two-seater, open Magna F2 were constructed during the same time period as the F1. The cars had six-cylinder overhead camshaft engines driven through a vertical dynamo with a capacity of 1271 cc generating 37 bhp at 4100 rpm. The gearbox was four-speed non-synchromesh and the centre-lock wire spoke wheels had 12-inch drums fitted with cable-operated brakes. Bore & Stroke: 57 × 83 mm.

16 *MG F3 Magna*

Like the F1 and F2, the third of the trio of Magna's 'F' types was made in the 1931–2 period—a total of 94 being produced. The six-cylinder overhead camshaft engine driven through a vertical dynamo had a capacity of 1271 cc generating 37 bhp at 4100 rpm. The gearbox was four-speed non-synchromesh and the centre-lock wire spoke wheels had 12-inch cable-operated brakes. Bore & Stroke: 57 × 83 mm.

17 *MG L1 Magna*

In all, 486 of the two-seater coupé and four-seater open and saloon L1s were made in the year from early 1933 to the start of 1934. The six-cylinder overhead camshaft engine driven through a vertical dynamo had a capacity of 1086 cc generating 41 bhp at 5500 rpm. The gearbox was four-speed non-synchromesh and the centre-lock wire spoke wheels had 12-inch cable-operated brakes. Bore & Stroke: 57 × 71 mm.

18 MG L2 Magna

A total of 90 models of the two-seater open L2 were produced in the same time period as the L1. The six-cylinder overhead camshaft engine driven through a vertical dynamo had a capacity of 1086 cc generating 41 bhp at 5500 rpm. The gearbox was four-speed non-synchromesh and the centre-spoke wire wheels had 12-inch cable-operated brakes. Bore & Stroke: 57 × 71 mm.

19 MG K1 (KA) Magnette

Approximately 54 of the four-seater K1 saloon with its KA engine containing three SU carburetters were made from the winter of 1932 to summer 1933. The six-cylinder overhead camshaft engine driven through a vertical dynamo had a capacity of 1086 cc generating 39 bhp at 5500 rpm. The gearbox was four-speed preselector and the centre-lock wire spoke wheels had 13-inch cable-operated brakes. Bore & Stroke: 57 × 71 mm.

20 MG K1 (KB) Magnette

74 models of this four-seater Tourer with its engine containing 2 SU carburetters were made in the same time period as the KB. The six-cylinder overhead camshaft engine driven through a vertical dynamo had a capacity of 1086 cc generating 41 bhp at 5500 rpm. The gearbox was four-speed non-synchromesh and the centre-lock wire spoke wheels had 13-inch cable-operated brakes. Bore & Stroke: 57 × 71mm.

21 MG K1 (KD) Magnette

A total of 53 four-seater open and four-seater saloons of the third of the K1 types were constructed during the summer of 1933 to spring of 1934. The six-cylinder overhead camshaft engine driven through a vertical dynamo had a capacity of 1271 cc generating 48.5 bhp at 5500 rpm. The gearbox was four-speed preselector and the centre-lock wire spoke wheels had 13-inch cable-operated brakes. Bore & Stroke: 57 × 83 mm.

22 MG K2 (KB) Magnette

A mere 16 examples of this two-seater open model were made during the summer of 1933 to early 1934. The six-cylinder overhead camshaft engine driven through a vertical dynamo had a capacity of 1086 cc generating 41 bhp at 5500 rpm. The gearbox was a four-speed non-synchromesh and the

centre-lock wire spoke wheels had 13-inch cable-operated brakes. Bore &
Stroke: 57 × 71 mm.

23   MG K2 (KD) Magnette

No more than five examples of this two-seater open model were made in the
closing months of 1933 and early 1934. The six-cylinder overhead camshaft
engine driven through a vertical dynamo had a capacity of 1271 cc generating
48.5 bhp at 5500 rpm. The gearbox was a four-speed preselector and the
centre-lock wire spoke wheels had 13-inch cable-operated brakes. Bore &
Stroke: 57 × 83 mm.

24   MG K3 Magnette

A total of 33 of the two-seater K3 Racing Magnette were built from late 1932
to the winter of 1934. The six-cylinder overhead camshaft engine driven
through a vertical dynamo had a capacity of 1086 cc generating 120 bhp at
6500 rpm. The gearbox was a four-speed preselector and the centre-lock wire
spoke wheels had 13-inch cable-operated brakes. Bore & Stroke: 57 × 71mm.

25   MG NA Magnette

Approximately 738 of the Magnette NA, which included the two-seater open
and four-seater open as well as the two-seater coupé, were constructed in the
two-year period from early 1934 to the winter of 1936. The six-cylinder
overhead camshaft engine driven through a vertical dynamo had a capacity
of 1271 cc generating 56 bhp at 5700 rpm. The gearbox was four-speed
non-synchromesh and the centre-lock wire spoke wheels had 12-inch cable-
operated brakes. Bore & Stroke: 57 × 83 mm.

26   MG NE Magnette

Only seven examples of this two-seater Racing model were constructed in
the winter of 1934. The six-cylinder overhead camshaft engine driven through
a vertical dynamo had a capacity of 1271 cc generating 74 bhp at 6500 rpm.
The gearbox was four-speed non-synchromesh and the centre-lock wire spoke
wheels had 12-inch cable-operated brakes. Bore & Stroke: 57 × 83 mm.

27   MG KN Magnette

The KN Magnette was made in a saloon version only, and a total of 201
were produced between the summer of 1934 and the winter of 1935. The
six-cylinder overhead camshaft engine driven through a vertical dynamo had
a capacity of 1271 cc generating 56 bhp at 5700 rpm. The gearbox was
four-speed non-synchromesh and the centre-lock wire spoke wheels had 13-
inch cable-operated brakes. Bore & Stroke: 57 × 83 mm.

28 MG PA Midget

The first in the new line of Midgets, the PA, which was manufactured in two-seater, four-seater and 'Airline' coupé versions, was in production from early 1934 to the summer of 1936. A total of 1973 models were constructed. The four-cylinder overhead camshaft engine driven through a vertical dynamo had a capacity of 847 cc generating 35 bhp at 5600 rpm. The gearbox was four-speed non-synchromesh and the centre-lock wire spoke wheels had 12-inch cable-operated brakes. Bore & Stroke: 57 × 83 mm.

29 MG PB Midget

About 526 of the two-seater and four-seater PB Midgets were made in the same 1934–6 period. The four-cylinder overhead camshaft engine driven through a vertical dynamo had a capacity of 939 cc generating 43 bhp at 5500 rpm. The gearbox was four-speed non-synchromesh and the centre-lock wire spoke wheels had 12-inch cable-operated brakes. Bore & Stroke: 60 × 83 mm.

30 MG QA Midget

Just eight of the two-seater QA Midgets were constructed during the summer and autumn of 1934. The four-cylinder overhead camshaft engine driven through a vertical dynamo had a capacity of 746 cc generating 113 bhp at 7200 rpm. The gearbox was a four-speed preselector with overload clutch, and the centre-lock wire spoke wheels had 12-inch cable-operated brakes. Bore & Stroke: 57 × 71 mm.

31 MG RA Midg '

Ten models of the single-seater RA Midget were built in the summer of 1935. The four-cylinder overhead camshaft engine was driven through a dummy vertical dynamo and had a capacity of 746 cc generating 113 bhp at 7200 rpm. The gearbox was a four-speed preselector with an overload clutch, and the centre-lock wire spoke wheels had 12-inch cable operated-brakes. Bore & Stroke: 57 × 71 mm.

32 MG TA Midget

The highest number of MGs manufactured to date—some 3,003 in all—of the two-seater, 'Airline' coupé and drophead coupé Midget TA came off the production line between the middle of 1936 and the spring of 1939. The four-cylinder pushrod overhead valve engine had a capacity of 1291 cc generating 52 bhp at 5000 rpm. The gearbox was a manual four-speed part-synchromesh, and the centre-lock wire spoke wheels had 9-inch Lockheed hydraulic brakes. Bore & Stroke: 63.5 × 102 mm.

33 MG TB Midget

A total of 379 of the two-seater and drophead coupé TB Midgets were produced during the year of 1939. The four-cylinder pushrod overhead valve engine had a capacity of 1250 cc generating 54 bhp at 5200 rpm. The gearbox was a part-synchromesh four-speed manual, and the centre-lock wire spoke wheels had 9-inch Lockheed hydraulic brakes. Bore & Stroke: 66.5 × 90 mm.

34 MG SA (2-Litre)

A total of 2,738 of the stylish saloon, four-seater Tourer and Tickford coupé versions of the SA were manufactured from the early months of 1936 to the outbreak of World War Two. The six-cylinder pushrod overhead valve engine had a capacity of 2280 cc (later models increased to 2322 cc) generating 75.3 bhp at 4300 rpm. The gearbox was a four-speed non-synchromesh on the early models (Synchromesh from late 1937), and the centre-lock wire spoke wheels had 12-inch Lockheed hydraulic brakes. Bore & Stroke: 69 × 102 mm (1936/7); 69.5 × 102 mm (1937/8).

35 MG VA (1½-Litre)

Slightly fewer VAs were built than the SA—a total of 2,407—during the summer of 1937 to late 1939, but the range of four-door saloons, four-seater tourers and Tickford coupés were equally stylish. The four-cylinder pushrod overhead valve engine had a capacity of 1548 cc generating 54 bhp at 4500 rpm. The gearbox was a part-synchromesh four-speed manual, and the centre-lock wire spoke wheels had 10-inch Lockheed hydraulic brakes. Bore & Stroke: 69.5 × 102 mm.

36 MG WA (2.6-Litre)

A total of 369 of the impressive WA four-door saloons, four-seater Tourers and Tickford coupés were manufactured during the winter of 1938 and the following year. The six-cylinder pushrod overhead valve engine had a capacity of 2561 cc generating 95.5 bhp at 4400 rpm. The gearbox was a part-synchromesh four-speed manual, and the centre-lock wire spoke wheels had 14-inch Lockheed hydraulic brakes. Bore & Stroke: 73 × 102 mm.

37 MG TC Midget

Approximately 10,000 of the two-seater open sports TC were built from late 1945 to the winter of 1949, when the post-War popularity of the MG really took off. The four-cylinder pushrod overhead valve engine had a capacity of 1250 cc generating 54.4 bhp at 5200 rpm. The gearbox was four-speed with synchromesh on 2, 3 and 4 gears, and the centre-lock wire spoke wheels had 9-inch Lockheed hydraulic brakes. Bore & Stroke: 66.5 × 90 mm.

38 *MG TD Midget*

The massive total of 29,664 models of the Mark I and Mark II TD Midget came from the MG Production Line from the autumn of 1949 to late in 1953. The four-cylinder pushrod overhead valve engine had a capacity of 1250 cc generating 54.4 bhp at 5200 rpm. (The Mark II increased this to 60 bhp.) The gearbox was four-speed with synchromesh on 2, 3 and 4 gears, and the bolt-on disc wheels had 9-inch Lockheed hydraulic brakes (2LS at front). Bore & Stroke: 66.5 × 90 mm.

39 *MG TF Midget*

Approximately 9,600 of the 1250 and 1500 versions of the two-seater TF were produced between the autumn of 1953 and the spring of 1955. The four-cylinder pushrod overhead valve engine generated 57 bhp at 5500 rpm on the 1250 cc; and 63 bhp at 5000 rpm on the 1466 cc. The gearbox was four-speed with synchromesh on 2, 3 and 4 gears, and the bolt-on disc wheels had 9-inch Lockheed hydraulic brakes (2LS on the front.) Bore & Stroke: 66.5 × 90 mm (1250 cc); 72 × 90 mm (1466 cc).

40 *MG 1¼ Litre Y-Type Saloon*

A total of 7,459 models of the YA and YB 1¼-litre four-door saloon were manufactured between the early months of 1947 and the winter of 1953. The four-cylinder pushrod overhead valve engine had a capacity of 1250 cc generating 46 bhp at 4800 rpm. The gearbox was four-speed with synchromesh on 2,3 and 4 gears, and the bolt-on disc wheels had 9-inch Lockheed hydraulic brakes (2LS on the front of the YB model.) Bore & Stroke: 66.5 × 90 mm.

41 *MG YT Tourer*

Some 877 models of this four-seater Tourer were constructed during the same period as the YA and YB saloons. The four-cylinder pushrod overhead valve engine had a capacity of 1250 cc generating 54 bhp at 5200 rpm. The gearbox was four-speed with synchromesh on 2, 3 and 4 gears, and the bolt-on disc wheels had 9-inch Lockheed hydraulic brakes. Bore & Stroke: 66.5 × 90 mm.

42 *MG ZA and ZB Magnette*

A total of 12,754 ZA Magnettes and 23,846 ZB Magnettes, both four-door saloons, were produced in the five-year period from 1953 to the end of 1958. The four-cylinder pushrod overhead valve engine powering these models had a capacity of 1489 cc generating 60 bhp at 4600 rpm on the ZA and 68.4 bhp at 5250 rpm on the ZB. Both models had four-speed gearboxes with synchromesh on 2, 3 and 4 gears, and the bolt-on disc wheels had 10-inch Lockheed hydraulic brakes (2LS on front). Bore & Stroke: 73 × 88.9 mm.

The sports car

with winning ways!

A captivating car to look at and a winner on the track, this new T.F. series M.G. Midget is all set to be the most popular of the breed! That longer and slightly lower bonnet houses an engine more vigorously alive than ever. And see what a lovely line the bowed radiator and streamlined headlamps give her! Your M.G. dealer will arrange a trial run.

Safety-glass is a standard M.G. feature.

NUFFIELD SERVICE IN EUROPE
Qualified M.G. owners planning a continental tour are invited to see their M.G. dealer for details of a free service to save foreign currency.

AGAIN — MG MIDGETS WIN TEAM AWARD
IN CIRCUIT OF IRELAND RALLY

The team prize in this major event, which covers a gruelling 1000-mile circuit, has now been awarded to MG teams at five of the last six meetings — a remarkable record of consistent success.

THE M.G. CAR COMPANY LIMITED, SALES DIVISION, COWLEY, OXFORD
London Showrooms: Stratton House, 80 Piccadilly, London, W.1
Overseas Business: Nuffield Exports Limited, Cowley, Oxford, and at 41 Piccadilly, London, W.1

1954: Sex appeal rears its head in this advertisement for the TF Series Midget.

43 MG Mark III Magnette

Approximately 15,676 of the Mark III Magnette saloon were manufactured between the spring of 1959 and the end of 1961. The four-cylinder pushrod overhead valve engine had a capacity of 1489 cc generating 66.5 bhp at 5200 rpm. The gearbox was four-speed synchromesh on 2, 3 and 4 gears, and the bolt-on disc wheels had 9-inch Girling hydraulic brakes (2LS on front). Bore & Stroke: 73 × 88.9 mm.

44 MG Mark IV Magnette

Almost as many Mark IV Magnette saloons were built as its predecessor in the years between 1961 and early 1968—a total of 13,738. The four-cylinder pushrod overhead valve engine had a capacity of 1622 cc generating 68 bhp at 5000 rpm. The gearbox was four-speed synchromesh on 2, 3 and 4 gears with—for the first time on an MG—optional 3-speed automatic. The bolt-on disc wheels had 9-inch Girling hydraulic brakes (2LS on front). Bore & Stroke: 76.2 × 88.9 mm.

45 MG 1100

Over 116,827 of the four-door 1100 saloon were manufactured between the autumn of 1962 and the middle of 1971. (A two-door saloon was also sold in the USA only). The four-cylinder (transverse) pushrod overhead valve engine had a capacity of 1098 cc generating 55 bhp at 5500 rpm. The gearbox was four-speed synchromesh on 2, 3 and four gears, with bolt-on disc wheels and Lockheed disc brakes at the front and drums at the rear. Bore & Stroke: 64.6 × 83.7 mm.

46 MG 1300

A total of 26,240 of the two- and four-door saloon were made by MG between late 1962 and the middle of 1971. The four-cylinder (transverse) pushrod overhead valve engine had a capacity of 1275 cc generating 70 bhp at 6000 rpm. The gearbox was a manual four-speed synchromesh, with a four-speed automatic gearbox an option from the end of 1967. The bolt-on disc wheels had Lockheed hydraulic disc brakes at the front and drums on the rear. Bore & Stroke: 70.6 × 81.3 mm.

47 MG MGA 1500

Built from the summer of 1955 to the spring of 1959, a total of 58,750 of the two-seater and fixed-head coupé MGA 1500 models were produced. The four-cylinder pushrod overhead valve engine had a capacity of 1489 cc generating 68 bhp at 5500 rpm. (This was improved on later models to 72 bhp at 5500 rpm.) The gearbox was four-speed synchromesh on 2, 3 and 4 gears, with bolt-on disc wheels with 10-inch Lockheed hydraulic brakes (2LS on front). Centre-lock wire spoke wheels were optional with this model. Bore & Stroke: 73× 88.9 mm.

48 *MG MGA Twin Cam*

A total of 2,111 of these two-seat tourers and two-seat coupés were made between the spring of 1958 and early 1960. The four-cylinder twin overhead camshaft chain-driven engine had a capacity of 1588 cc generating 108 bhp at 6700 rpm. The gear box was four-speed synchromesh on 2, 3 and 4 gears, and the centre-lock perforated disc wheels had Dunlop disc brakes at the front and rear. Bore and Stroke: 75.4 × 88.9 mm.

49 *MG MGA 1600*

A grand total of 40,220 of the Mark I and Mark II 1600 two-seat Tourers and two-seat coupés were produced from the spring of 1959 until the summer of 1962. The four-cylinder pushrod overhead valve engine had a capacity on the Mark I of 1588 cc and on the Mark II of 1622 cc, generating, respectively, 80 bhp at 5600 rpm and 93 bhp at 5500 rpm. Both had four-speed gearboxes with synchromesh on 2, 3 and 4 gears, and the bolt-on disc wheels had Lockheed hydraulic discs on the front and drums on the rear. Bore & Stroke: 75.4 × 88.9 mm (Mark I); 76.2 × 88.9 mm (Mark II).

50 *MG MGB Mark I*

The MGB Mark I was produced from the winter of 1962 to late 1967, with some 115,898 of the GHN. 3 two-seater completed, plus 21,835 of the GHD.3 GT coupé, which was introduced in late 1965. The four-cylinder pushrod overhead valve engine had a capacity of 1789 cc generating 95 bhp at 5400 rpm. The gear box was a part-synchromesh four-speed manual, and the bolt-on disc wheels (with optional centre-lock wire wheels) had Lockheed hydraulic discs at the front and drums at the rear. Bore and Stroke: 80.3 × 88.9 mm.

51 *MG MGB Mark II*

The Mark II two-seater Tourers and GT coupés were produced from the autumn of 1967 until late in 1980—a total of 375,147 being made. The four-cylinder pushrod overhead valve engine had a capacity of 1798 cc generating 95 bhp at 5400 rpm. The gearbox was an all-synchromesh four-speed manual; with automatic an optional extra from 1967 to 1973. The wheels were 'Rostyle' bolt-on, and the car had Lockheed disc brakes at the front and drums at the rear. Bore & Stroke: 80.3 × 88.9 mm.

52 *MG MGC*

The MGC was in production from the autumn of 1967 for two years and

For a really brilliant performance book a seat in the proven MGB or MGB GT

I would like to test-drive an MGB.........
MGB GT.........
(tick preference)

I have for part-exchange a

(make).....................(model).............

(year).................... (mileage)...........

Name...

Address..

.............................Tel:.................

A demonstration and a test-drive of either of these well-tried cars will convince you of their outstanding quality.

The MGB, internationally famous and built in the true sporting tradition, has 1798 cc power. So has the MGB GT, built in the elegant style of a genuine Grand Tourer and a truly classic performer.

Henlys have a specialised export sales division to help you with any purchase problems. Contact us today.

HENLYS
Friars Road, Ipswich, Tel: 75431

1970: The international fame of the marque is used to promote the MGB.

during this time a total of 4,542 of the two-seater Tourer version and 4,457 of the GT coupé were made. The six-cylinder pushrod overhead valve engine had a capacity of 2912 cc generating 145 bhp at 5250 rpm. The gearbox was all-synchromesh four-speed manual (automatic was optional), and the bolt-on disc wheels had Lockheed hydraulic discs at the front and drums at the rear, plus a servo-assisted system. Bore & Stroke: 83.4 × 88.9 mm.

53 *MG MGB GT V8*
Some 2,591 of these GT coupé models were manufactured between the winter of 1972 and the summer of 1976. The eight-cylinder pushrod overhead valve engine had a capacity of 3528 cc generating 137 bhp at 5000 rpm. The gearbox was an all-synchromesh four-speed manual with overdrive on top; and the Dunlop bolt-on wheels had Lockheed disc brakes at the front and drums at the rear, plus a servo-assisted system. Bore & Stroke: 88.9 × 71.1 mm.

54 *MG Midget Mark I*
The first of the 'modern' two-seater open Tourer Mark I Midgets was produced in two versions, the GAN 1 and GAN 2, between the summer of 1961 and the spring of 1964. Some 16,080 of the GAN 1 were made, and 9,601 of the GAN 2. The four-cylinder pushrod overhead valve engine had a capacity of 948 cc in the GAN 1, generating 46.4 bhp at 5500 rpm; and 1098 cc in the GAN 2, generating 55 bhp at 5500 rpm. The cars had a four-speed gearbox with synchromesh on 2, 3 and 4 gears; and the bolt-on disc wheels had Lockheed hydraulic brakes, with 2LS on the front of the GAN 1 and discs on the front of the GAN 2. Bore & Stroke: 62.9 × 76.2 (GAN1); 64.6 × 83.7 (GAN 2).

55 *MG Midget Mark II*
A total of 26,601 of the Mark II GAN 3 two-seater open Tourer were made between early 1964 and the winter of 1966. The four-cylinder pushrod overhead valve engine had a capacity of 1098 cc generating 59 bhp at 5750 rpm. The gearbox was four-speed with synchromesh on 2, 3 and 4 gears, and the bolt-on disc wheels had Lockheed hydraulic discs on the front and drums on the rear. Bore & Stroke: 64.6 × 83.7 mm.

56 *MG Midget Mark III*
Between late 1966 and the end of 1974, a total of 13,722 of the GAN 4 Mark III and 86,650 of the GAN 5 Mark III (the latter of which began production in late 1969) were made by MG. The four-cylinder pushrod overhead valve engine had a capacity of 1275 cc generating 65 bhp at 6000 rpm. The gearbox was four-speed with synchromesh on 2, 3 and 4 gears, and the bolt-on disc

wheels had Lockheed disc brakes at the front and drums on the rear. Bore & Stroke: 70.6 × 81.3 mm.

57 *MG Midget Mark III (GAN 6)*

A total of 72,185 of the two-seater open Tourer GAN 6 Mark III were produced from the autumn of 1974 to late in 1979, when the MG Octagon disappeared from sports cars. The four-cylinder pushrod overhead valve engine had a capacity of 1491 cc generating 65 bhp at 5500 rpm. The gearbox was a four-speed synchromesh manual, and the 'Rostyle' bolt-on wheels had Lockheed hydraulic discs at the front and drums on the rear. Bore & Stroke: 73.7 × 87.4 mm.

58 *MG RV8*

The MG sports car has returned in a sumptuous two-seater that is seen as a natural evolution from the vastly successful MGB originally launched 30 years ago. The new MG RV8 has a Rover 3.9-litre eight-cylinder engine with a capacity of 3946 cc generating 190 PS at 4750 rpm. The gearbox is a five-speed manual, and the bolt-on wheels are cast alloy of lattice spoke design with ventilated discs on the front and 9-inch drums on the rear, with a servo-assisted system. Bore & Stroke: 94.0 × 71.1 mm.

NOTE: This listing does not include the MG saloon car variants that were made in the Eighties, i.e. the Metro (in production from spring 1982 onwards), the Metro Turbo (from summer 1982), the Maestro (only made from the spring of 1983 to winter 1984) and the Montego (spring 1984 onwards).

The latest MG, the RV8, promoted in the traditional way with a touch of style and glamour. This official Rover photograph was taken at Hartwell House Hotel in the Vale of Aylesbury.

The MG File

1   Places of Interest

The Morris Garages Showrooms, 36/37, Queen Street, Oxford
The showrooms where Cecil Kimber worked as sales manager from 1921. A year later he became general manager and began the development which would result in the MG cars. The upper floor is currently a restaurant.

Morris Garage, Longwall Street, Oxford
Here the early Bullnose Morris's were built and Cecil Kimber created his famous 'Old Number One' which won the Gold Medal in the 1925 Land's End Trial.

Bainton Road MG Factory, Oxford
MG's were built here from 1925 in the two bays which can still be seen on the left-hand side through the gate in Bainton Road.

Edmond Road Factory, Oxford
Here the MG 14/40, 18/80 and 'M' type Midgets were built between 1927 to 1929 until lack of space forced the company to move to nearby Abingdon.

MG Car Company, Marcham Road, Abingdon
The expanding MG company moved here in 1929 to occupy a site purchased from the Pavlova Leather Company. In this factory, Cecil Kimber and his dedicated workforce created the legendary models, and it remained in operation until closed down by British Leyland in 1980. Many of the original buildings are still standing, including the administration block or 'Top Office' occupied by Cecil Kimber and his successor, John Thornley.

The National Motor Museum, Beaulieu, Hampshire
The country's most famous motor museum near the South Coast contains examples of the best known MG models which are always on display and regularly featured in the historical Cavalcades of cars.

A 1935 MG 'P' type recently photographed outside the entrance to the MG
Car Company Works in Abingdon. (The RoadRunner Collection)

The Heritage Motor Centre, Banbury Road, Gaydon, Warwickshire

Over 300 historic British cars are on display at Gaydon, including several of the most famous MGs—'Old Number One' and the two great record breakers; EX 135, in which 'Goldie' Gardner exceeded 200 mph, and Stirling Moss's EX-181. Plus the very first MG RV8 off the production line.

Brooklands Race Track, Weybridge, Surrey

This world-famous 'Birthplace of Motor Sport', opened in 1907, was the venue where MGs broke several records in the years before its closure in 1946. Reopened, its museum has several MGs. It is the starting point each May for the annual MG Regency Run to Brighton—the marque's own version of the famous RAC London-to-Brighton Veteran Car Rally.

2 *The M.G. Clubs*

The MG Car Club Ltd, PO Box 251, Kimber House, Abingdon, Oxford OX14 1FF

01235 555552

The oldest of the MG enthusiast clubs, having been founded in 1931, and now embracing a network of Registers each specialising in the various models. The club publishes its own magazine, *Safety Fast*, and through its Centres is responsible for a large number of competitions and rallies held throughout the country. It also has affiliated clubs all over the world, whose addresses can be obtained on application to Kimber House.

A rare photograph taken inside MG's Abingdon works in 1930 during the production of 18/80 cars and 'M' type Midgets. (Early MG Society)

The MG Owners' Club, Octagon House, Swavesey, Cambridge, CB4 5QZ

Claimed to be 'The World's Largest One-Make Motor Club' with a membership of over 50,000, this organisation is geared to the post-war MG models and runs a highly efficient service for members, including repairs at its own workshop, recommending spares and insurance, as well as co-ordinating meetings and rallies around the country. It also publishes a monthly magazine, *Enjoying MG*.

Octagon Car Club, 36, Queensville Avenue, Stafford ST17 4LS

Formed by a small group of enthusiasts in the Stafford and Stoke-on-Trent area, the club caters for pre-1956 MGs from the 14/80 up to the TF and YB models. The club runs social and competitive events and publishes a monthly magazine.

The Early MG Society, Kimber House, Hirston Lane, Storrington, W. Sussex

The youngest and smallest of the new clubs, this society looks specifically after the interests of the 14 and 18 hp vintage MGs. It publishes information and advice for members, is busy creating a fascinating archive of press and photographic material, and arranges social meetings and rallies.

3 Selected Reading

Allison, Mike, *MG: The Magic of the Marque* (Autostyle Publications, 1989).
Aspden, Richard, *The Classic MG* (Bookthrift, 1986).
Autocar, *MG Sports Cars* (Bay View Books, 1990).
Blower, W. E., *MG Workshop Manual* (Haynes Group, 1980).
Clarke, R., *MG TC 1945–1949.* (Brooklands Books, 1986).
Clausager, Anders, *MG: The Book of the Car* (Smithmark Publishing, 1983).
Clausager, Anders, *Original MG 'T' Series* (Motorbooks Int., 1989).
Fletcher, Rivers, *MG Past and Present* (Foulis, 1985).
Gardner, Bill, *MG: One and a Half Litre* (Gardner, 1991).
Garnier, Peter (Ed), *MG Sports Cars* (Hamlyn, 1979).
Harvey, Chris, *The Immortal 'T' Series* (Foulis, 1989).
Jennings, P. L., *Early MG* (Addendum Publishing, 1985).
Knudson, Richard, *MG International* (Motor Racing Publications, 1977).
Knudson, Richard, *Illustrated MG Buyer's Guide* (Motorbooks International, 1983).
Knudson, Richard, *MG: The Sports Car America Loved First* (Motorcars Unlimited, 1975).
Knudson, Richard (ed.), *The Cecil Kimber Book* (The 'T' Register of New England, 1988).
Labban, Brian, *MGB: The Complete Story* (Crowood Press, 1990).
MacLennan, John, *MG: The Art of Abingdon* (Motor Racing Publications, 1982).
Thornley, John W., *Maintaining the Breed* (Motor Racing Publications, 1950).
Tyler, Jim, *MG Midget* (Osprey, 1992).
McComb, F. Wilson, *MG by McComb* (Osprey, 1978).
McComb, F. Wilson, *MGA 1500, 1600 Twin Camb* (Osprey, 1983).
Penberthy, Ian, *MG* (Mallard Books, 1991).
Philip, George, *Restoration MG* (Haynes Group, 1992).
Porter, Lindsay, *MGB* (Foulis, 1989).
Porter, Lindsay, *MG Midget* (Foulis, 1989).
Robson, Graham, *The MGA, MGB and MGC: A Collector's Guide* (Motorbooks Int., 1982).
Robson, Graham, *The 'T' Series MG* (Motorbooks Int., 1980).
Ullyett, Kenneth, *The MG Companion* (Stanley Paul, 1960).

(Opposite) The Test Hill at Brooklands—all eyes are on Rivers Fletcher driving his 1930 Double Twelve Midget with co-owner Elizabeth Wigg up this demanding incline. (Andrew Roberts)

Vitrikas, Robert P., *MGA: A History and Restoration Guide* (Scarborough Faire Inc., 1992).

Wherry, Joseph H., *MG: The Sports Car Supreme, 1924–1980* (Oak Tree Publications, 1982).

Wood, Jonathan & Burrell, Lionel, *MGB: The Illustrated History* (Foulis, 1988).

MG Magic—fun, excitement and real motoring in two of the best sports cars in the world! (Ron Cover)